# PRACTICAL BATTERY DESIGN AND CONTROL

For a complete listing of titles in the
*Artech House Power Engineering Library,*
turn to the back of this book.

# PRACTICAL BATTERY DESIGN AND CONTROL

Naoki Matsumura

BOSTON | LONDON
artechhouse.com

**Library of Congress Cataloging-in-Publication Data**
A catalog record for this book is available from the U.S. Library of Congress.

**British Library Cataloguing in Publication Data**
A catalogue record for this book is available from the British Library.

**Cover design by Andy Meaden Creative**

ISBN 13: 978-1-63081-975-0


**Artech House**
**685 Canton Street**
**Norwood, MA 02062**



10 9 8 7 6 5 4 3 2 1

# CONTENTS

# 2
# APPLICATION OF ELECTROCHEMISTRY TO BATTERIES 19



# 3
# BATTERY IMPEDANCE AND ITS IMPACT ON BATTERY LIFE 35

# 8

## FUEL CELL 159



# 9

## OTHER BATTERY-RELATED TECHNOLOGIES 177

# FOREWORD

Naoki Matsumura has provided a technical work that provides the reader with a clear path for learning about batteries and associated requirements. Naoki is a recognized expert in the field and has used his knowledge to write a comprehensive work that gives the reader a foundational understanding of the common-use cases and parameters to successfully deploy batteries. This is a reference book that provides the theory and how-to for both the student and those that want to expand their knowledge. As a neophyte in the battery field, I found this book to be readable despite the technical depth, and will absolutely keep this on my desk as a reference.

*Marshall Smith*
*Senior Director, Supply Chain Enabling Technologies*
*Intel Corporation*
*March 2023*

# PREFACE

No battery, no life. These days, energy storage systems (i.e., batteries in this book) are used everywhere, from laptop PCs, smartphones, electric vehicles, to backup batteries. In the future, even airplanes may be electrified with batteries. Without batteries, it is difficult to live our lives.

As the battery market grew, demand for battery engineering jobs increased worldwide. To work in the industry, a wide variety of knowledge is now required, ranging from electrochemistry to material science, physical chemistry, and machine learning. While there are books and university classes that focus on each topic, there is not a book that compiles necessary theories and equations from all of these topics and describes their practical applications to solve real-life battery problems. For example, when you are asked to make a 1-Ah battery, how do you calculate the grams of cathode and anode materials required? When an industrial designer of a new laptop PC allocates dimensions for a battery, how does one determine the expected battery life? How do you predict battery longevity and extend it with machine-learning algorithms? When an electric vehicle has a solar panel on its roof, how many additional miles can it provide? When you research a post Li-ion battery, what is the expected battery voltage and battery life of the new chemistry? When you are asked to use a supercapacitor instead of a Li-ion battery, what are the advantages and disadvantages? The list goes on.

This book helps readers to answer these questions with intuitive explanations and practical problem sets and answers. The contents are designed for battery engineers to expand their knowledge and advance their career to the next level. Furthermore, it supports anyone, including college students who are interested in battery technologies to obtain practical knowledge. My sincere hope is that readers will contribute to the betterment of the world with what they learned from this book.

# ACKNOWLEDGMENTS

I would like to express my sincere appreciation to Intel Corporation, especially Marshall Smith, for his support in publishing this book. I also appreciate my colleagues and mentors who greatly influenced my career: Andy Keates, Ramon Cancel, Ron Woodbeck, Morgan Hartnell, Vivek Ramani, Brian Fritz, Tod F. Schiff, and Chris Capener.

Last but not least, I would like to thank my wife, Akiko Matsumura, and daughter, Otoha Matsumura, for understanding the importance of publishing this book and offering substantial support.

# 1

# LI-ION BATTERY OVERVIEW AND SPEC

## 1.1 INTRODUCTION: BATTERY HISTORY TO LI-ION BATTERY

Humankind invented and commercialized the lithium-ion (Li-ion) battery, a rechargeable battery with the best energy density both volumetrically and gravimetrically. It has been a long journey to get there. Battery history starts in 1791 when Luigi Galvani discovered the basis of the battery principle from electricity in frog legs [1] and Alessandro Volta invented the voltaic battery. Over the next 180 years, researchers continuously made improvements, enhancing energy density by inventing novel batteries, such as lead-acid batteries, nickel-cadmium batteries, and nickel-metal hydride batteries. Finally in the 1970s and 1980s, technologies to realize Li-ion batteries were invented by John B. Goodenough, M. Stanley Whittingham, and Akira Yoshino. They were awarded the Nobel Prize in Chemistry 2019 [2]. Their invention opened the door to enable small-sized and long-lasting electronic devices.

## 1.2 STRUCTURE OF THE LI-ION BATTERY

Today's mobile phones and laptop personal computers (PCs) typically use Li-ion batteries. The battery consists of cathode, anode, and separator sheets as shown in Figure 1.1. In the battery, the sheets of cathode, anode, and separator are stacked, wound, or folded, and inserted

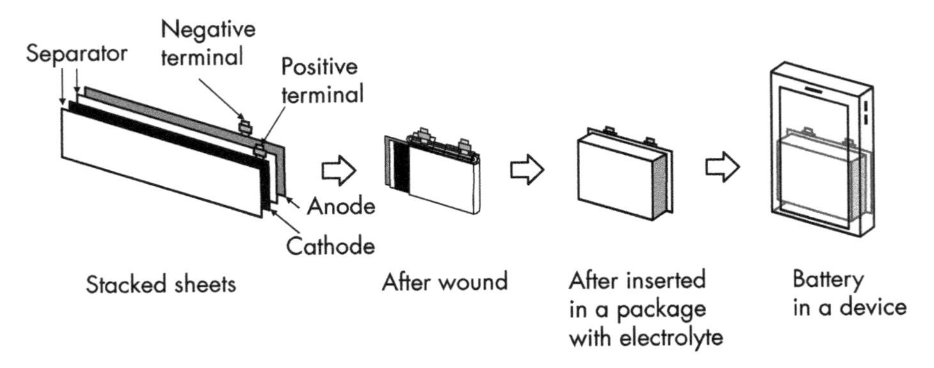


**Figure 1.1** An example of Li-ion battery cell structure.

in a package with an electrolyte. Such a cathode and anode assembly is called a cell.

The cathode works as the positive electrode (+) and the anode works as the negative electrode (–). The separator is a porous sheet that allows lithium ions to pass between cathode and anode during charge and discharge. The separator is electronically nonconductive and prevents cathode-anode contact that would cause an internal short circuit and in the worst case, may result in an explosion. A liquid or polymer electrolyte allows the lithium ions to move between cathode and anode.

## 1.3 INTUITIVE UNDERSTANDING OF CHARGING/DISCHARGING MECHANISMS

### 1.3.1 Charging Mechanism

In a Li-ion battery, the cathode active material is typically a lithium metal oxide, such as $LiCoO_2$, $Li(Ni_xMn_yCo_z)O_2$, and $Li(Ni_xCo_yAl_z)O_2$. The anode active material is typically graphite. Figure 1.2 is a schematic illustration of the cathode and anode before and during charge.

Both $LiCoO_2$ and graphite have a layered structure as shown in Figure 1.2. The electrolyte contains $LiPF_6$, which exists as lithium ions ($Li^+$) and hexafluorophosphate ions ($PF_6^-$).

When the battery is empty, lithium is stored in cathode as part of layered $LiCoO_2$ structure. During charge, lithium is electrochemically extracted from the cathode by the charger, releases an electron and becomes a lithium ion. Electrons move to the anode through the

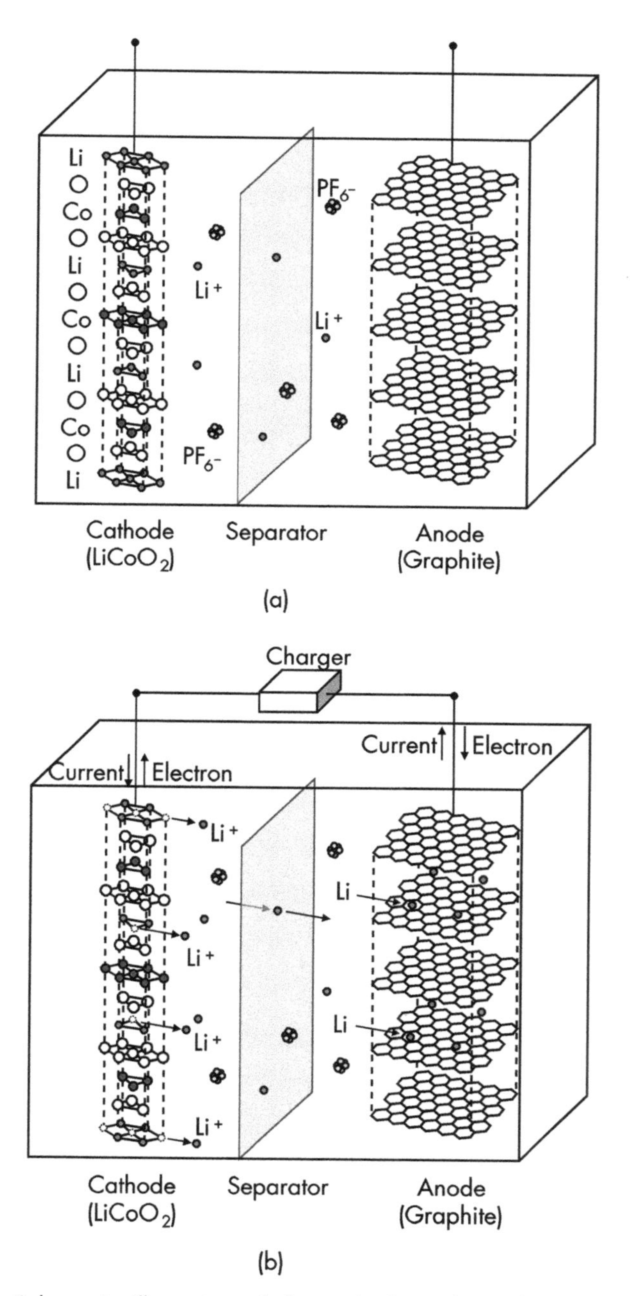


**Figure 1.2** Schematic illustration of the cathode and anode in a Li-ion battery: (a) before charge, and (b) during charge.

charger that is an external circuit. Lithium ions move towards the anode through the electrolyte inside the cell. Lithium ions and electrons meet again at the anode, where the lithium is stored in between graphite layers as shown in Figure 1.2(b), and the graphite is said to be lithiated. The driving force to move lithium ions and electrons is the charger. The charger provides external energy to the cell in this way.

### 1.3.2 Discharging Mechanism

During discharge, the reverse reactions to charge happen. Lithium is electrochemically extracted (i.e., deintercalated) from the anode, releases an electron, and becomes a lithium ion. Electrons move to the cathode through the outer system (e.g., laptop PC and smartphone). Then, lithium ions and electrons meet again in the cathode as shown in Figure 1.2(a). The driving force to move lithium ions and electrons is the difference of chemical potentials between the cathode and the anode. The chemical potential difference leads to battery voltage.

### 1.3.3 Chemical Reactions During Charge and Discharge

The chemical reactions during charge and discharge can be expressed as follows.

Half reaction at the anode:

$$\mathrm{LiC_6} = \mathrm{C_6} + \mathrm{Li^+} + \mathrm{e^-} \tag{1.1}$$

where $LiC_6$ is lithiated graphite and $C_6$ is graphite.

After charge, lithium in the $LiC_6$ anode would like to move back to the cathode side and become part of $LiCoO_2$. This is because lithium is more stable in the cathode due to a chemical potential difference. More details about chemical potentials can be found in Chapter 2. Once the outer system (e.g., laptop PC and smartphone) starts using the energy of the battery that is discharge, lithium in $LiC_6$ becomes lithium ion ($Li^+$), releases an electron ($e^-$), and moves to the cathode side.

Half reaction at the cathode:

$$\mathrm{Li_{1-\mathit{x}}CoO_2} + x\mathrm{Li^+} + x\mathrm{e^-} = \mathrm{LiCoO_2} \tag{1.2}$$

where x is the utilization rate of lithium in $LiCoO_2$ ($0 \leq x \leq 1$).

During discharge, lithium ions come to $Li_{1-x}CoO_2$ and receive electrons. In the cathode reaction, practically speaking, $x$ cannot be 1 because too much extraction of lithium from $LiCoO_2$ causes collapse of the layered crystal structure, resulting in poor rechargeability (i.e., poor cycle life). In practice, the limit of $x$ depends on technology generations (e.g., $x = 0.65$). Efforts to increase $x$ continue to be made by researchers, including investigations of additives into $LiCoO_2$ to make the structure more robust.

A combination of half reactions gives the following full reaction:

$$Li_{1-x}CoO_2 + xLiC_6 = LiCoO_2 + xC_6 \tag{1.3}$$

There are several other ways to express the full reaction of Li-ion batteries. In this book, (1.3) is chosen because it is the most practical, explaining mass balance of cathode and anode.

*Redox Reaction*

In general, chemical reactions of batteries are called redox reactions that consist of a reduction reaction and an oxidation reaction. The combination of reduction and oxidation is redox. In (1.1), when discharging, $LiC_6$ loses an electron and is categorized in oxidation by definition. In (1.2), when discharging, $Li_{1-x}CoO_2$ gains an electron and is categorized in reduction by definition.

For the readers of this book who are focused on practical usage, it is most important to theoretically and intuitively understand how ions and electrons move during charge and discharge.

## 1.4 KEY INNOVATIONS TO REALIZE LI-ION BATTERY

This section explains the key innovations in the history of Li-ion batteries [2]. Lithium is one of the elements that is reactive, thus can release an electron with high energy. By utilizing lithium, Stanley Whittingham invented the functional Li-ion battery with a titanium disulfide cathode ($TiS_2$) and a lithium metal anode in the early 1970s.

In 1980, John Goodenough discovered that lithium cobalt oxide ($LiCoO_2$) works as a cathode and doubled battery voltage from ~2V to ~4V. This is because $LiCoO_2$ has a strong chemical affinity between lithium and metal oxide. Lithium cobalt oxide is still used in batteries for systems that need the highest energy density, such as laptop PCs and smartphones.

While Li-ion batteries provide high-energy density, eventually it was discovered that using lithium metal as the anode can cause safety risks. During charge, lithium comes from the cathode and is plated on the anode as lithium metal. Such plating does not grow flat but generates dendrite (needles of lithium metal) as shown in Figure 1.3.

When a dendrite grows, it may eventually go through the separator, reach the cathode, and cause an internal short circuit which, in the worst case, may result in high-heat generation and/or explosion.

To address the dendrite challenge, Akira Yoshino chose petroleum coke as the anode instead of lithium metal as it can store lithium safely in between carbon layers during charge. This worked well and resulted in Li-ion batteries without dangerous lithium metal.

In 1991, the Li-ion battery was commercialized by Sony [3], led by Yoshio Nishi. Since that time, it has been further improved by many researchers and engineers in many areas, such as chemistry, manufacturing processes, and battery management system.

## 1.5 NECESSARY BATTERY KNOWLEDGE TO READ A BATTERY SPECIFICATION

To work in the battery-related industries, it is important to understand battery terminologies.

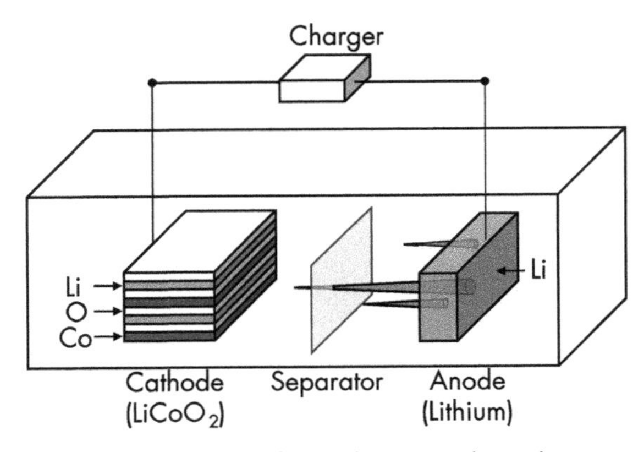


**Figure 1.3** An example of dendrite from lithium metal anode.

A battery specification sheet (or spec) and a battery label on a battery pack show a lot of information, for example, 3.8V, 2716 mAh (2.716 Ah), and 10.35 Wh, as shown by the label in Figure 1.4.

Table 1.1 is another example that is part of a battery specification. It lists 0.5C CC to 4.35V, then CV at 4.35V until 0.02C and Max discharging current: 1.0C.

This section explains the necessary knowledge to read the battery spec and label.

### 1.5.1 Basic Terminologies

The first step is to review the basic terminologies.

*Ampere (A)* is the unit of current that is electron flow. 1A equals $\sim 6.24 \times 10^{18}$ electrons/second, meaning that $\sim 6.24 \times 10^{18}$ electrons flow per second where $\sim 6.24 \times 10^{18}$ electrons are defined as 1 coulomb (1C) of charge, just like we call 12 pieces as one dozen. Therefore, 1A equals 1 C/second.

*Voltage (V)* is the unit of electron potential. 1V equals 1 joule/coulomb. This means that, if 1 coulomb of electric charge has 1V, it has 1 joule of energy.

*Watt (W)* is the unit of power that equals V × A. As V × A equals (joule/coulomb) × (coulomb/sec) = (joule/sec), watt is energy per second.

Typical analogy for these terminologies is a water tank as shown in Figure 1.5.

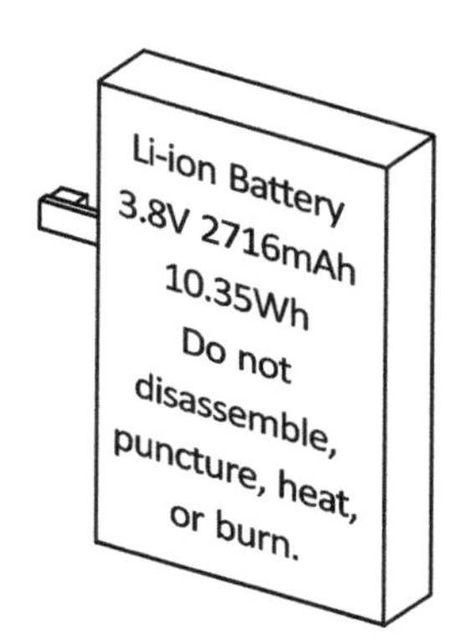


**Figure 1.4** An example of battery label.

**Table 1.1**
An Example for Part of a Battery Specification

| No. | Item | Specification |
|---|---|---|
| 1 | Minimum capacity | 4800 mAh at 0.2C discharge (3.0V cutoff) |
| 2 | Nominal voltage | 3.80V |
| 3 | Standard charging method | 0.5C CC to 4.35V, then CV at 4.35V until 0.02C |
| 4 | Maximum charging current | 1.0C at CC |
| 5 | Maximum discharging current | 1.0C to 3.0V |
| 6 | Operation temperature for charging | 0°C to 45°C |
| 7 | Operation temperature for discharging | –20°C to 60°C |
| 8 | Initial impedance | < 40 mohm at 50% SOC (AC 1kHz) |
| 9 | Cycle life | At least 80% recoverable capacity after 500 cycles of 0.5C CC-CV charging, terminating at 0.05C, and 0.5C discharging at 25°C |
| 10 | Storage life | At least 80% recoverable capacity after 12 months storage with 50% SOC at 23°C |
| | | At least 50% recoverable capacity after 3 months storage with 50% SOC at 40°C |
| 11 | Thickness before shipping | Maximum 3.74 mm |
| 12 | Thickness after 500 cycles | Maximum 4.00 mm |
| 13 | Width | Maximum 73.0 mm |
| 14 | Length | Maximum 110.0 mm |

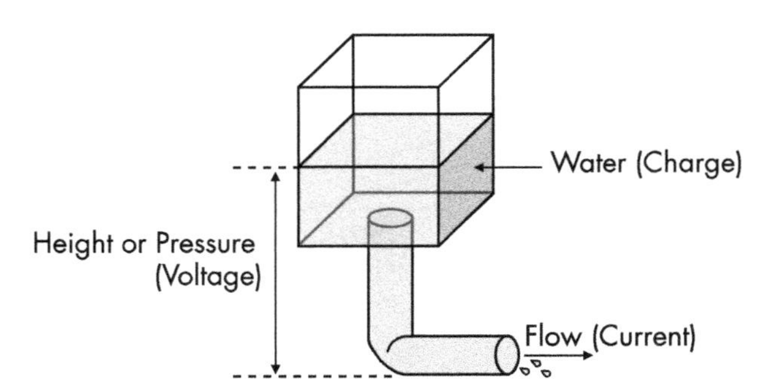


**Figure 1.5** Water tank as an analogy of ampere, voltage, and watt.

Electron or battery charge is water, current is water flow, voltage is height, which gives water pressure, and a battery is the tank. A higher voltage battery is analogous to a higher tank position that gives more energy.

***Exercises***

1. If a battery can supply 1A at 3.8V, how much power can it provide?
2. How much energy can the battery in the question above provide per second?
3. If a battery provides 3.8W at 2A, what is the battery voltage?
4. If a battery provides 3.8W at 1.9V, what is the supply current?
5. How many electrons are flowing per second in case of 10A? (1 coulomb = $6.24 \times 10^{18}$ electrons.)

***Answers***

1. 1A × 3.8V = 3.8W.
2. W is joule/sec. 3.8W provides 3.8 joule per second.
3. W = V × I. This can be transformed as V = W/I. Therefore, 3.8W / 2A = 1.9V.
4. W = V × I. This can be transformed as I = W/V. Therefore, 3.8W/1.9V = 2.0A.
5. Ampere (A) is coulomb/sec. 10A is 10 coulombs/sec. As 1 coulomb ≈ $6.24 \times 10^{18}$ electrons, 10A is ~$6.24 \times 10^{19}$ electrons/sec.

* * *

### 1.5.2 Battery Terminologies

The next step is to fully understand the battery specific terminologies.

*Ampere-hour (Ah)* is the capacity or amount of electric charge. 1 Ah is the capacity where 1 ampere continues for 1 hour, meaning 1A × 1h = 1 A × h = 1 Ah. Do not pronounce Ah as "/a/." It is pronounced as "ampere hour."

***Exercises***

1. How many hours can a 1-Ah battery last at 0.5A?
2. How many hours can a 5-Ah battery last at 0.1A?

***Answers***

1. 1 Ah / 0.5A = 1 ~~A~~h / 0.5~~A~~ = 2 hours.

2. 5 Ah / 0.1A = 5 ~~A~~h / 0.1~~A~~ = 50 hours.

Note that impedance is not considered in this exercise and is explained in Chapter 3.

* * *

*Watt-hour (Wh),* is the amount of energy and is calculated as Ah × (nominal battery voltage). 1 Wh is the energy where 1W continues for 1h, meaning 1W × 1h = 1 W×h = 1 Wh. Wh is pronounced as "watt hour."

***Exercises***

1. How many hours can a 1-Wh battery last at 0.5W?
2. How many hours can a 5-Wh battery last at 0.1W?
3. What is Ah of a 11.4-Wh battery with nominal 3.8V?

***Answers***

1. 1 Wh / 0.5W = 1 ~~W~~h / 0.5~~W~~ = 2 hours.
2. 5 Wh / 0.1W = 5 ~~W~~h / 0.1~~W~~ = 50 hours.
3. 11.4 Wh / 3.8V = 11.4 (V×A×h) / 3.8V = 3 Ah.

Again, impedance is not considered in this exercise and is explained in Chapter 3.

* * *

*C-rate* is the rated current relative to the full-charge capacity of the battery. C is pronounced as the letter C "/si:/." 1C is the current to discharge the fully charged battery in 1 hour. For example, 1C for a 1-Ah battery is 1A. 0.5C for a 1-Ah battery is 0.5A. 0.5C for a 2-Ah battery is 1A.

***Exercise***

1. What is 2C for a 2-Ah battery?

***Answer***

1. 2C × 2 Ah = 4A.

* * *

Note that C-rate is always calculated from the full-charge Ah in the spec, regardless of the battery state of charge (SOC). For example, for a 2-Ah battery that is half charged and 1 Ah is stored, 1C is 2A. For a 2-Ah battery that is empty, 1C is still 2A. In the battery-related industries, C-rate is often used, especially in the spec. This is because C-rate gives the same stress to the battery regardless of the capacity of the battery. Also, battery manufacturers can easily specify the recommended charging current, max discharge current, and so forth. For example, when a battery manufacturer typically supports up to 1C discharge, a 1-Ah battery from the company is capable of 1A discharge. A 2-Ah battery from the company is capable of 2A discharge.

*SOC* is the battery charge percent relative to the full-charge capacity. For a 1-Ah battery, 50% SOC is 0.5 Ah. For a 2-Ah battery, 30% SOC is 0.6 Ah.

***Exercise***

1. When the SOC of a 2-Ah battery is 60%, what is the capacity?

***Answer***

1. 2 Ah $\times$ 0.6 = 1.2 Ah.

* * *

### 1.5.3 Battery Charging Spec

For a Li-ion battery, CC-CV charging is used in general. CC and CV are the abbreviations of constant current and constant voltage, respectively. Figure 1.6 is an example of CC-CV charging.

This figure explains how current, voltage, and SOC change during charge. It starts with 0% SOC where the battery is empty. CC-CV charging starts with CC charging, which is constant current charging. This is shown as step 1 in Figure 1.6. CC for Li-ion batteries in laptop PCs or smartphones is typically performed at a C-rate of 0.5C to 1C. As CC charging continues, battery voltage and SOC go up. When the battery voltage reaches charge cutoff voltage, which is defined in the battery spec, charging mode shifts from CC to CV as shown in step 2 in Figure 1.6. Charge cutoff voltage is a fixed value that depends on anode/cathode chemistry and technology generation. Batteries on the market today typically define this cutoff voltage somewhere between 4.2V to 4.4V per cell. During CV charging, current decreases naturally as the battery capacity gets filled. When current reaches charge cutoff

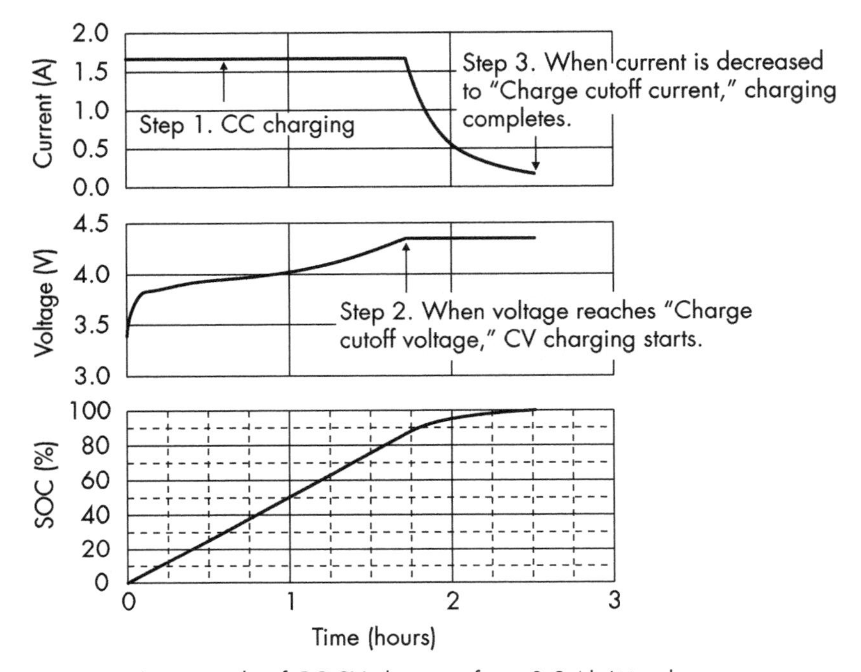


**Figure 1.6** An example of CC-CV charging for a 3.3-Ah Li-ion battery.

current, such as 0.02C to 0.05C, battery charging completes. Key specs during charge are (1) current at CC stage, (2) charge cutoff voltage to switch from CC to CV, and (3) cutoff current during CV charging. Battery charging needs to be performed carefully as written in the spec. This is because deviation from the spec, such as overcharging, may result in harmful situations.

***Exercise***

1. Refer to Figure 1.6. You have a 3.3-Ah battery that is empty.
   a. How many hours would it take to complete a 50% charge at 0.5C current?
   b. How many hours would it take to complete a 75% charge at 1.65A?
   c. If CV charging is stopped at ~0.50A cutoff, how much charge percentage can be roughly expected and how many hours will it take to charge from empty?

***Answer***

a. It takes 1 hour: 50% of 3.3 Ah is 1.65 Ah. Charging was performed at 0.5C constant current. Thus 0.5C of a 3.3-Ah battery is 1.65A, and 1.65 Ah/1.65A is 1 hour. Another solution is that 0.5C is the current that discharges 50% of the battery capacity in 1 hour. This can also be applied to CC charging. Therefore, the answer is 1 hour.

b. It takes 1.5 hours: 75% of 3.3 Ah is 2.475 Ah. Charging was performed at 0.5C constant current that is 1.65A, and 2.475 Ah/1.65A = 1.5 hours.

c. As the dotted line in Figure 1.7 shows, when current is decreased at ~0.5A during CV charging, time is 2 hours and SOC indicates around 95%.

* * *

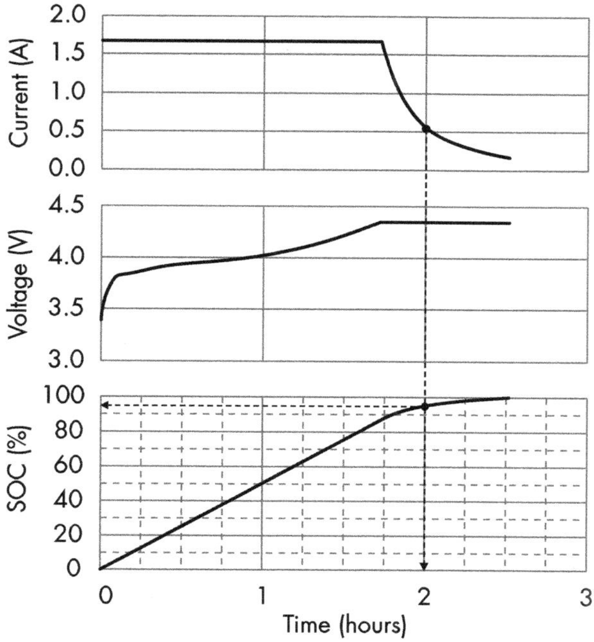

**Figure 1.7** CC-CV charging for a 3.3-Ah Li-ion battery indicating ~0.5A CV cutoff.

### 1.5.4 Battery Cycle Life and Storage Life Spec

Battery full-charge capacity naturally decreases after multiple charging and discharging cycles. It also decreases after storage. For example, when the fresh battery of a new smartphone provides 10 hours of video playback time with full charge, the time will then decrease after repeated charging and discharging because of battery degradation.

*Cycle life* is the battery spec or data that shows how much capacity is recoverable after charging and discharging cycles. One example spec is "At least 80% recoverable capacity after 500 cycles of 0.5C CC-CV charging, terminating at 0.05C, and 0.5C discharging at 25°C." This means that when the battery is charged at 0.5C CC with a CV termination current of 0.05C, and discharged at 0.5C in the environment at 25°C, at least 80% capacity is recoverable after 500 charging and discharging cycles.

***Exercise***

1. For a 1-Ah cell with the same cycle life spec as above, how much is the full charge capacity after 500 cycles of 0.5C CC-CV charge and 0.5C discharge at 25°C?

***Answer***

1. 1 Ah $\times$ 0.8 = 0.8 Ah. Full charge capacity is at least 0.8 Ah.

* * *

*Shelf-life or storage-life* is the spec or data that shows how much capacity is recoverable after the battery is stored at a certain condition for a certain period. The following is an example:

"At least 80% recoverable capacity after 12 months storage with 50% SOC at 23°C." This means that when the battery is charged to 50% SOC and stored at 23°C for 12 months, at least 80% of the original capacity is recoverable when the battery is fully charged.

Other examples: "At least 50% recoverable capacity after 3 months storage with 50% SOC at 40°C" and "At least 90% recoverable capacity after 1 month storage with 100% SOC at 23°C." As shelf life depends on battery SOC and storage temperature, battery specs typically show several shelf-life specs.

***Exercise***

1. When a 1-Ah cell is charged to 50% SOC and stored at 40°C for 3 months, how much is the full charge capacity? Refer to the abovementioned spec.

***Answer***

1. The spec shows 50% recoverable capacity on this storage condition. Thus, 1 Ah × 0.5 = 0.5 Ah. Therefore, full charge capacity is at least 0.5 Ah.

* * *

In addition, there are several other key items in the spec. For example, item numbers 4 and 5 in Table 1.1 show maximun 1.0C for charging and discharging current. If the cell supports a higher C-rate, user experience will be better because of faster charging speed and higher discharging capability, which leads to more powerful response from a system (e.g., laptop PC). However, it is a trade-off against the cell capacity. The details are explained in Chapters 4 and 11. Item numbers 6 and 7 show operating temperature for charging and discharging. These are important specs for safe battery operation, which is covered in Chapter 6. Item number 8 shows a battery impedance spec. Battery impedance largely affects the system performance such as battery life. The details are explained in Chapter 3. Battery dimensions in item numbers 11 through 14 are also important. When a cell packaging is soft, the cell swells after charging/discharging cycles because of the nature of the degradation. Some cell specifications show the dimensions, especially thickness after swelling (e.g., item number 12 in Table 1.1). The details of swelling are explained in Chapter 6.

## 1.6 SUMMARY

In this chapter, we learned the following:

- How lithium ions and electrons of a Li-ion battery move during charge and discharge.
- Battery terminologies and how to read a battery spec.
- How to calculate battery life and charge time using C-rate.

## 1.7 PROBLEMS

Problem 1.1

1. Write half reactions at cathode and anode of a Li-ion battery with $LiCoO_2$ cathode and $C_6$ anode. Illustrate how lithium ion, electron, and current flows between cathode and anode during charge. Also explain what drives the reaction during charge.

Hint: Cathode starts as $LiCoO_2$, anode starts as $C_6$, and $x$ lithium moves from cathode to anode as $xLi^+$.

2. When a Li-ion battery consists of $LiNiO_2$ cathode that has a similar layered structure to $LiCoO_2$ and $C_6$ anode, write half reactions at cathode and anode during charge.

Hint: Cathode starts as $LiNiO_2$ and forms $Li_{1-x}NiO_2$. Anode starts as $C_6$ and forms $LiC_6$.

Answer 1.1

1. Cathode: $LiCoO_2 = Li_{1-x}CoO_2 + xLi^+ + xe^-$
   Anode: $C_6 + Li^+ + e^- = LiC_6$
   Figure 1.2(b) shows the flow of lithium ion, electron, and current during charge. An external charger drives the charging reaction.
2. Cathode: $LiNiO_2 = Li_{1-x}NiO_2 + xLi^+ + xe^-$
   Anode: $C_6 + Li^+ + e^- = LiC_6$

Problem 1.2

Refer to Figure 1.6. When you have a 3.3-Ah battery that is empty,

1. What do 1C and 0.5C mean for a 3.3-Ah cell?

2. How many hours would it take to complete a 40% charge at 0.5C current?

3. How many hours would it take to complete a 40% charge at 0.4C current?

4. If charging completes at the end of CC charge, how much charge percentage can be expected, and how many hours will it take to charge from empty? Read the figure.

Answer 1.2

1. 1C means 3.3A, and 0.5C means 1.65A.
2. 0.8 hours, 0.4/0.5C = 0.8h.

3. 1 hour, 0.4/0.4C = 1h.

4. Figure 1.6 shows that the end of CC charging corresponds to slightly less than 90% charge and slightly less than 1.75 hours.

Problem 1.3

You received a battery spec of a new smartphone. It shows 15.2 Wh, nom. 3.8V, 2C-CC-CV.

1. What is the capacity of this battery?

2. If 2C-CC-CV means CC charging at 2C, followed by CV charging, what is the charging current in amperes?

3. If constant current continues from 0% SOC to 80% SOC, how many hours does it take for 80% charge?

4. How many coulombs are needed for 80% charge?

Answer 1.3

1. 15.2 Wh / 3.8V = 4.0 Ah.

2. 2C of 4.0 Ah is 8.0A.

3. 80% of 4.0 Ah is 3.2 Ah. 3.2 Ah charging at 8.0A takes 3.2 Ah/8.0A = 0.4h.

4. 8A is 8 coulomb/sec. 0.4h is 3600 sec × 0.4h = 1440 seconds. Thus, 8 coulomb/sec × 1440 sec = 11520 coulombs.

Problem 1.4

You have a smartphone that uses a Li-ion battery. When you charge it from empty, it takes 1.5 hours to achieve 75%. But it takes another 1 hour to charge the remaining 25%, which is a slower charging rate. Explain the reason.

Answer 1.4

CC-CV charging is typically used for Li-ion batteries. It starts with CC charging, followed by CV charging. When charging is switched from CC to CV, charging current decreases. This is why charging slows down in the later phase of charging.

Problem 1.5

Refer to Figure 1.6. In Problem 1.2.2, it is calculated that an empty 3.3-Ah battery takes 0.8h to charge 40%. When the battery is at

50% SOC, does additional 40% charging to the 50% charged battery take the same 0.8h? If not, explain the reason.

Answer 1.5

According to Figure 1.6, it takes slightly longer than 0.8h to charge an additional 40% from 50% SOC to 90% SOC. This is because battery charging enters the CV step, which is slower than the CC charging. Chapter 4 explains the details of charging.

## References

[1] Galvani, Luigi, L., "De viribus electricitatis in motu musculari. Commentarius," *De Bonoiensi Scientiarum et Artium Intituo atque Academie Commentarii,* Vol. 7, 1791, pp. 363–418.

[2] The Royal Swedish Academy of Sciences, "Popular Science Background: They Developed the World's Most Powerful Battery," https://www.nobelprize.org/uploads/2019/10/popular-chemistryprize2019.pdf.

[3] Nishi, Y., "Lithium Ion Secondary Batteries: Past 10 Years and the Future," *Journal of Power Sources,* Vol. 100, No. 1–2, 2001, pp. 101–106.

# 2

# APPLICATION OF ELECTROCHEMISTRY TO BATTERIES

## 2.1 INTRODUCTION

In the previous chapter, we learned how to read a battery spec, which depends on battery chemistries, manufacturers, models, and so forth. This chapter explains why the nominal Li-ion battery voltage is ~3.7V, how to calculate the battery voltage of different chemistries, and how many grams of Li-ion battery materials are needed to run a smartphone for 10 hours.

## 2.2 BATTERY VOLTAGE SCIENCE AND APPLICATION

### 2.2.1 Li-ion Battery Voltage

Figure 2.1 shows how Li-ion battery voltage changes during 0.2C discharge.

From the fully charged state, battery voltage decreases during discharge, and battery charge level goes down. Safe discharge cutoff voltage of a Li-ion battery cell depends on the spec and is typically 3.0V/cell in the case of a Li-ion battery with $LiCoO_2$ cathode and graphite anode. In this figure, discharging completes when the

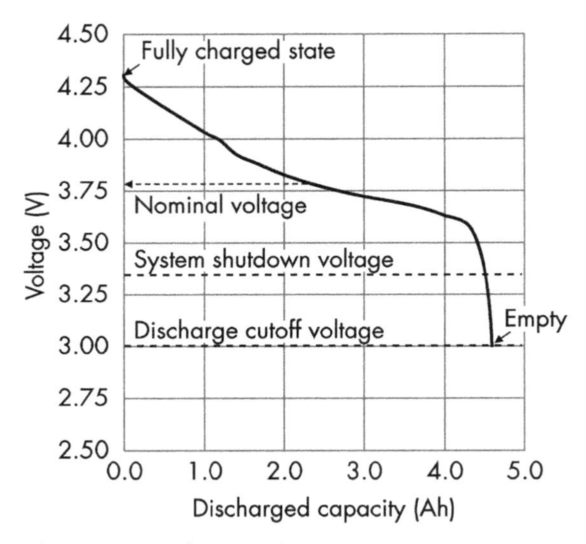


**Figure 2.1** Discharge curve of a Li-ion battery at 25°C.

battery voltage reaches 3.0V, and around 4.6 Ah is provided. When a battery is used in a system, such as a smartphone, the system usually has a different shutdown voltage from the discharge cutoff voltage. This is, for example, due to minimum voltage requirement from integrated circuits (ICs) or voltage regulators in the system. If the system shutdown voltage is higher than the battery discharge cutoff voltage, the system shuts down when the battery voltage reaches the system shutdown voltage. In this case, some battery capacity is left unused or reserved for minimal system functions such as maintaining the internal clock.

Nominal voltage is an averaged voltage calculated by Wh/Ah. The nominal voltage of a single-cell Li-ion battery is typically 3.7V to 3.8V in the case of $LiCoO_2$ cathode and graphite anode. It is determined by laws of nature associated with the choice of cathode and anode chemistry.

### 2.2.2 Energy Level Difference

The full reaction of a Li-ion battery with $LiCoO_2$ cathode and graphite anode is expressed as follows:

$$\mathrm{Li}_{1-x}\mathrm{CoO}_2 + x\mathrm{LiC}_6 = \mathrm{LiCoO}_2 + x\mathrm{C}_6 \tag{2.1}$$

To understand the natural equilibrium of (2.1) and how much energy can be derived, we can look at a quantity called the Gibbs free energy. Each substance has its own Gibbs free energy that is defined by the following formula:

$$G = G^{\circ} + RT \ln a \tag{2.2}$$

where $G^{\circ}$, pronounced as $G$-naught, is the Gibbs free energy at standard condition, which is at 25°C and 1 atm, $R$ is the gas constant, which is ~8.314 J · mol$^{-1}$ · K$^{-1}$, $T$ is temperature in Kelvin, $a$ is activity. Activity is 1 at standard condition (thus, $RT \ln a = 0$) and decreases/increases depending on the state of substance, such as purity for solid, concentration for ideal solutions, and partial pressure for gases. The Gibbs free energy per mole is also called the chemical potential.

$Li_{1-x}CoO_2$ has its own Gibbs free energy, $G(Li_{1-x}CoO_2)$. $LiC_6$, $LiCoO_2$, and $C_6$ also have their own Gibbs free energy, which are $G(LiC_6)$, $G(LiCoO_2)$, and $G(LiC_6)$, respectively. Gibbs free energy, $G$, is calculated via $G = H - TS$ where $H$ is enthalpy, $T$ is temperature in Kelvin, and $S$ is entropy. $H$ and $S$ are typically determined using calorimetry or theoretical simulations such as first-principles calculations.

The difference of the Gibbs free energy between the right side and left side of the chemical reaction is released or absorbed when the chemical reaction proceeds. In (2.1), the difference can be expressed as:

$$\Delta G = G\left(LiCoO_2\right) + xG\left(C_6\right) - G\left(Li_{1-x}CoO_2\right) - xG\left(LiC_6\right) \tag{2.3}$$

If $\Delta G$ is negative, the reaction is spontaneous, meaning that the right side of (2.1) is more stable. If $\Delta G$ is positive, the reaction is not naturally possible, meaning that the left side of the reaction is more stable.

Figure 2.2 shows the schematic illustration of the Gibbs free energy difference between the right side and left side of (2.1).

The left side of (2.1) is at the charged state. The right side of the equation is at the discharged state. As the Gibbs free energy of the discharged state is lower and more stable than that of the charged state, $\Delta G$ is negative. Therefore, the battery substances naturally change from the charged state to the discharged state, releasing energy.

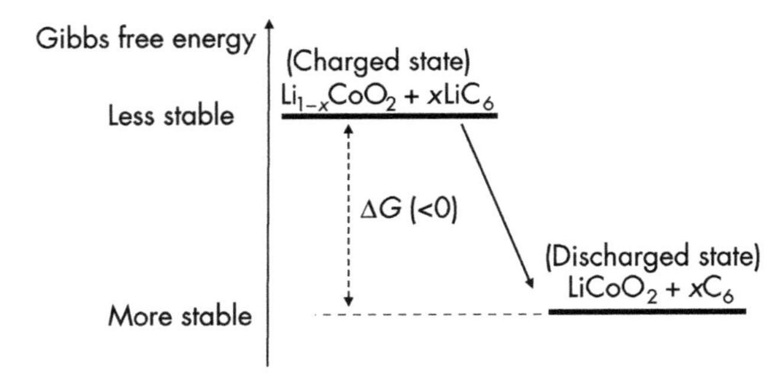


**Figure 2.2** Schematic illustration of Gibbs free energy difference.

### 2.2.3 Nernst Equation and Application

What does $\Delta G$ practically mean to energy storage systems? $\Delta G$ leads to electromotive force (emf), which is battery voltage via the Nernst equation:

$$\Delta G = -nFE \tag{2.4}$$

where $n$ is the number of electrons transferred in the reaction, $F$ is the Faraday constant, which is ~96485 C/mol, and $E$ is the cell potential, which is the voltage of the battery.

***Exercise***

There is a battery that consists of hydrogen and copper as shown in Figure 2.3. When the chemical reaction of the battery is expressed as follows and $\Delta G$ is –65.0 kJ, calculate the theoretical battery voltage:

$$H_2(g) + Cu^{2+}(aq) = 2H^+(aq) + Cu \tag{2.5}$$

*Hint:* $H_2$ releases 2 electrons. $Cu^{2+}$ receives 2 electrons.

***Answer***

With the Nernst equation, $-65.0 \times 10^3 = -2 \times 96485 \times E$. Solving the equation gives that the theoretical battery voltage, $E$, is ~0.337V.

In reality, the observed voltage is different from the theoretical voltage. The activity of substances changes during the chemical

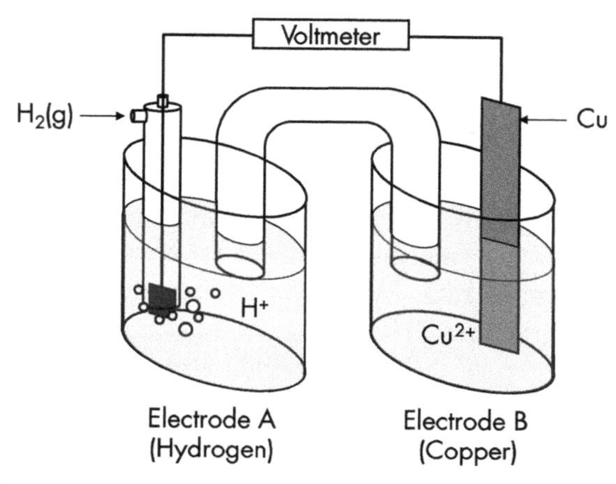


**Figure 2.3** Illustration of a battery with copper and hydrogen.

reaction because of concentration change, pressure change, and so forth. For example, when hydrogen pressure and hydrogen ion concentration in (2.5) changes during the reaction, $G(H_2(g))$ and $G(H^+(aq))$ change because activity, $a$, in (2.2) changes. This also changes $\Delta G$ of (2.5), resulting in a different voltage from the theoretical one.

* * *

### 2.2.4 Standard Potential of Half Reaction

In Section 1.3, we learned that chemical reactions of batteries are redox reactions where the reduction reaction and the oxidation reaction occur simultaneously. Equation (2.5) is also a redox reaction where the following half reactions occur:

$$\text{Electrode A: } H_2(g) = 2H^+ + 2e^- \tag{2.6}$$

$$\text{Electrode B: } Cu^{2+} + 2e^- = Cu \tag{2.7}$$

Electrode A is anode and the oxidation side by definition as it loses electrons. Electrode B is cathode and the reduction side as it gains electrons. When the theoretical voltage is measured in Figure 2.3, all substances are at the standard condition where temperature is 25°C and activity of the Gibbs free energy is 1. By definition, when the activity is 1, the gas is under 1 atm, concentration of the solutes is 1 mol/l and pure substances are used.

When electrode B is replaced with a different substance such as zinc, the half reaction is $Zn^{2+} + 2e^- = Zn$ and −0.76V is observed. Then, what is the voltage when electrode B is lithium, magnesium, or other substances? To answer the question, researchers have measured voltages of different substances by standardizing electrode A as the standard hydrogen electrode (SHE) and replacing substances in electrode B. As the voltage is also called potential, the summary of the voltages at the reduction side is named as standard reduction potentials. Table 2.1 is an example of standard reduction potentials [1].

As all potentials are relative to the hydrogen electrode, the hydrogen reaction is zero volts. In the table, the lower substances prefer to move to the right side of the reaction by obtaining electrons. This is because the right side is more stable. The upper substances prefer to move to the left side by releasing electrons. This is because the left side is more stable. Lithium metal is least stable in the table and prefers to become lithium ion by releasing an electron.

With the table, it is possible to calculate voltages of substances when electrode A is changed. For example, when electrode A is changed from the standard hydrogen electrode to zinc and electrode B is copper, half and full reactions are as follows:

$$\text{Electrode A: } Zn = Zn^{2+} + 2e^- \quad +0.76V \tag{2.8}$$

**Table 2.1**
An Example of Standard Reduction Potentials (*From:* [1])

| | Substances | Potential (V vs. SHE*) |
|---|---|---|
| Prefer to move to the left side ↑ | $Li^+ + e^- = Li$ | -3.04 |
| | $Mg^{2+} + 2e^- = Mg$ | -2.36 |
| | $Zn^{2+} + 2e^- = Zn$ | -0.76 |
| | $2H^+ + 2e^- = H_2$ | 0.00 (SHE*) |
| Prefer to move to the right side ↓ | $Cu^{2+} + 2e^- = Cu$ | +0.34 |
| | $Au^{3+} + 3e^- = Au$ | +1.50 |

* SHE: Standard Hydrogen Electrode

$$\text{Electrode B: } Cu^{2+} + 2e^{-} = Cu \quad +0.34V \tag{2.9}$$

$$\text{Full Reaction: } Zn + Cu^{2+} = Zn^{2+} + Cu \tag{2.10}$$

Equation (2.8) is intentionally reversed from the one in Table 2.1. This is to align the number of electrons to (2.9) and cancel them in the full reaction of (2.10). As (2.8) is reversed, the voltage is also reversed from −0.76V to +0.76V. By combining (2.8) and (2.9) to form (2.10), the voltages of (2.8) and (2.9) are added together as 0.76V + 0.34V = +1.10V. This is the voltage of (2.10).

### 2.2.5 Li-ion Battery Voltage Science

When electrode A is lithium, what is the voltage? To answer the question, let's think about the case where electrode A is lithium and electrode B is copper. The half reaction of the electrode B is the same as (2.9). The half reaction of lithium in electrode A is expressed as follows:

$$\text{Electrode A: } Li = Li^{+} + e^{-} \quad +3.04V \tag{2.11}$$

This equation is reversed from the one in Table 2.1. To combine it with (2.9) and cancel the electrons, (2.11) requires two electrons. Multiplying (2.11) by two gives the following equation:

$$\text{Electrode A: } 2Li = 2Li^{+} + 2e^{-} \quad +3.04V \tag{2.12}$$

Note that the voltage in (2.12) stays the same and is not multiplied by two. This is because voltage is not influenced by the quantity of the substances in the reaction, while $\Delta G$ of (2.12), $\Delta G$(2.12), is influenced by the quantity and is doubled from $\Delta G$ of (2.11), $\Delta G$(2.11). For example, when the Nernst equation is applied to (2.12), the following equation is derived:

$$\Delta G(2.12) = -2 \times F \times E \tag{2.13}$$

where $E$ is the voltage of (2.12).

When the Nernst equation is applied to (2.11), the following equation is derived:

$$\Delta G(2.11) = -1 \times F \times E' \qquad (2.14)$$

where $E'$ is the voltage of (2.11).

As $\Delta G(2.12) = 2 \times \Delta G(2.11)$, (2.13) and (2.14) lead to $E = E'$.

By combining (2.9) and (2.12), the voltage with lithium in electrode A and copper in electrode B is as follows:

$$2\text{Li} + \text{Cu}^{2+} = 2\text{Li}^{+} + \text{Cu} \quad +3.38V \qquad (2.15)$$

In other words, copper voltage is 3.38V versus $Li/Li^+$, while it is also true that copper voltage is 0.34V versus SHE. It depends on what the voltage of the substance is relative to.

In designing Li-ion batteries, it is often conventional to reference the voltages of both the cathode and anode relative to $Li/Li^+$ rather than the SHE. Then, what is the voltage of the chemistries used in Li-ion batteries relative to $Li/Li^+$? Figure 2.4 shows a schematic illustration of $LiCoO_2$ and graphite voltages relative to lithium. For consumer electronics devices, $LiCoO_2$ and graphite are major cathode and anode, respectively, because of high-energy density.

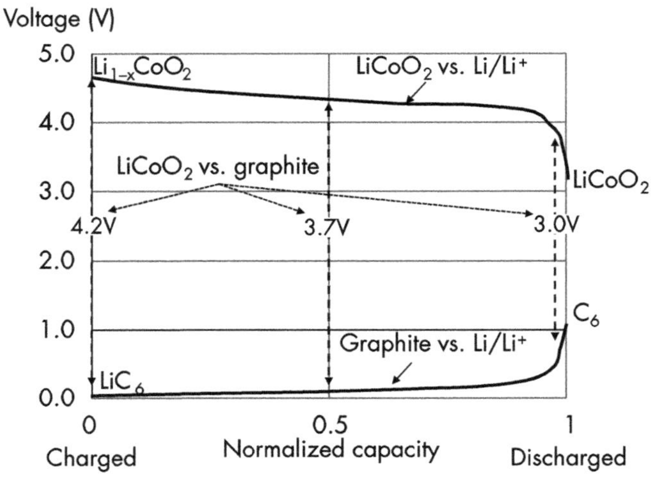


**Figure 2.4** Schematic illustration of $LiCoO_2$ and graphite voltages relative to $Li/Li^+$.

In Figure 2.4, the leftmost is the charged state and the rightmost is the discharged state. The voltage of $LiCoO_2$ is relative to $Li/Li^+$, starting as $Li_{1-x}CoO_2$ and ending as $LiCoO_2$. The voltage of graphite is also relative to $Li/Li^+$, starting as $LiC_6$ and ending as $C_6$ through several intermediate Li-C stages [2]. So what is the Li-ion battery voltage with $LiCoO_2$ cathode and graphite anode? The answer is the difference between $LiCoO_2$ versus $Li/Li^+$ and graphite versus $Li/Li^+$. This is because the battery voltage is the relative difference in voltage between the cathode and the anode. Such battery voltages are shown as dotted lines in Figure 2.4. The battery voltage starts at ~4.2V, ends at 3.0V, and the middle voltage, which is nominal voltage, is 3.7V. This is the reason why the charge cutoff voltage of a Li-ion battery is ~4.2V, the discharge cutoff voltage is 3.0V, and the nominal voltage is ~3.7V. Battery voltage changes during charge and discharge because the Gibbs free energy of $Li_{1-x}CoO_2$ and graphite changes with the SOC.

### 2.2.6 Voltage of Future Batteries

There are many options for the cathode and the anode of Li-ion batteries. Figure 2.5 is the summary of key materials with capacity (mAh/cc) in *x*-axis and nominal voltage versus $Li/Li^+$ in *y*-axis [3–17].

Some materials, especially metal anodes, swell when charged, or lithiate. Such materials require additional room to compensate for swelling. For practical comparison, the capacity in the *x*-axis of

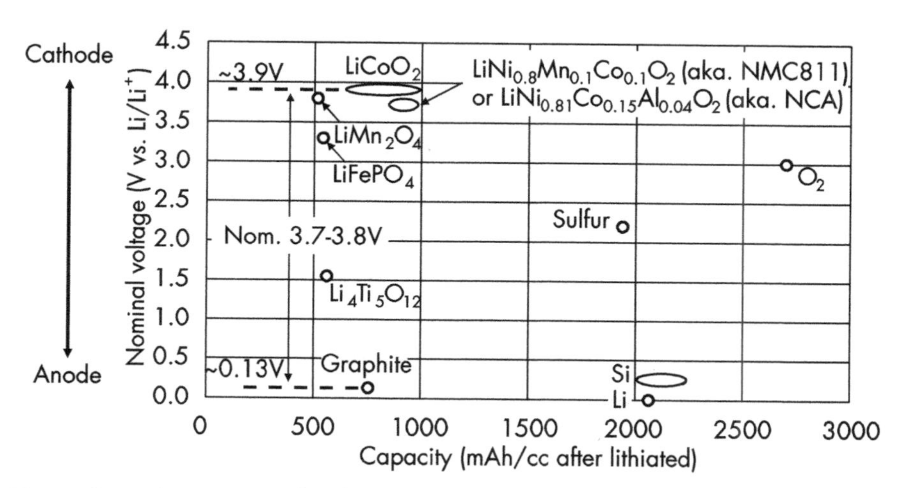


**Figure 2.5** Comparison of cathode and anode materials for Li-ion batteries [3–17].

Figure 2.5 is calculated based on the volume after lithiation. In the figure, high-voltage materials are cathode candidates, low-voltage materials are anode candidates, and medium-voltage materials can be both cathode and anode depending on the combination with the other material. When a battery is made with $LiCoO_2$ cathode and graphite anode, because $LiCoO_2$ is at ~3.9V versus $Li/Li^+$ and graphite is at ~0.13V versus $Li/Li^+$, the battery provides nominal 3.7V to 3.8V by the difference between cathode and anode voltages. It is also possible to estimate future battery voltage by knowing the nominal voltage of the material.

***Exercise***

When sulfur with ~2.2V versus $Li/Li^+$ is used as cathode and lithium with 0V versus $Li/Li^+$ is used as anode, what is the battery voltage?

***Answer***

The battery is expected to provide $2.2 - 0 = 2.2$V as nominal voltage.

On a side note, a lithium-sulfur battery is actively researched because of its high-energy density and low cost, compared to today's Li-ion battery with $LiCoO_2$ and graphite.

## 2.3 APPLICATION OF ELECTROCHEMISTRY TO BATTERY DESIGN

### 2.3.1 Faraday's Law of Electrolysis

When the reaction, $Li = Li^+ + e^-$, occurs, one lithium releases one electron. If there is one mole of lithium, then 1 mole of electrons is released. 1 mole of electrons can be converted to ~96485 coulombs, which is called the Faraday constant. The symbol C is used for coulombs. In Chapter 1, we already learned that 1A equals 1 C/second. Let's apply the knowledge for a practical use case.

***Exercise***

1A comes from a lithium electrode for 1 second by the reaction of $Li = Li^+ + e^-$.

1. How many moles of lithium are needed at least? Use 96485 C/mol for the Faraday constant.

2. How many grams of lithium are needed at least? Use 6.94 g/mol for the molar mass of lithium.

***Answer***

1. To provide 1A (i.e., 1 C/1 sec) for one second, 1C electron is needed by the calculation of 1 C/1 sec × 1 sec. Thus, 1C is 1/96485 mol. The ratio of lithium to electrons is 1:1. Therefore, 1/96485 mol is needed at least.
2. 6.94 g/mol × 1/96485 mol ≈ $7.19 \times 10^{-5}$ g. Therefore, ~$7.19 \times 10^{-5}$ g is needed at least.

* * *

Then, when the electrode is changed from lithium to magnesium that releases two electrons by the reaction of $Mg = Mg^{2+} + 2e^{-}$, how does the calculation change to provide 1A for 1 second? The required number of electrons does not change from 1/96485 mol. However, the ratio of magnesium to electrons is 1:2. Therefore, (1/2) × (1/96485 mol) of magnesium is needed.

This section explained the method to calculate the required mass of materials from the required number of electrons. This is what Michel Faraday discovered and is now called Faraday's law of electrolysis.

### 2.3.2 Amount of Cathode and Anode Needed

It's time to demonstrate the obtained knowledge in real Li-ion batteries.

***Exercise***

As a battery engineer in an imaginary X Corporation, you are involved in a new smartphone design that is planning to use a Li-ion rechargeable battery. Half reactions of the battery are as follows:

$$\text{Anode: } LiC_6 = Li^{+} + e^{-} + C_6 \tag{2.16}$$

$$\text{Cathode: } Li_{0.35}CoO_2 + 0.65Li^{+} + 0.65e^{-} = LiCoO_2 \tag{2.17}$$

Your peer in the marketing team requires 10 hours of video play time. Your peer in the electrical engineering team informed you that the new smartphone needs 0.76W on average during video play. The battery is nominal 3.8V. When $C_6$ is 72.07 g/mol, $LiCoO_2$ is 97.87 g/mol, and the Faraday constant is 96485 C/mol, answer the following questions:

1. How many Ah are required from the battery at least?
2. How many grams of $LiCoO_2$ and graphite ($C_6$) are required at least?

***Answer***

1. 10 hours at 0.76W requires 0.76W × 10h = 7.6 Wh battery. When nominal voltage is 3.8V, the battery has 7.6 Wh/3.8V = 2 Ah.
2. 2 Ah means 2 coulombs/sec for 1 hour, which is 2 C/sec × 3600 sec = 7200C. To provide 2 Ah, 7200C of electrons are required. Thus 7200C equals 7200/96485 mol electrons.

   Equation (2.16) shows that the ratio of electrons to $C_6$ is 1:1. Therefore, the required $C_6$ is 7200/96485 mol × 72.07 g/mol ≈ 5.378g.Equation (2.17) shows that the ratio of electrons to $LiCoO_2$ is 0.65:1. Therefore, the required $LiCoO_2$ is 7200/96485 mol × 97.87 g/mol × 1/0.65 ≈ 11.24g.

* * *

In reality, slightly more graphite is used to safely store all the possible lithium from the cathode in the graphite structure after charging. Otherwise, any excess lithium may generate dendrites on the graphite, which creates a safety risk. Thoughtful engineers also consider battery impedance impact on battery life, which is explained in Chapter 3.

On a side note, Chapter 1 explained that the full reaction with $LiCoO_2$ and graphite is $Li_{1-x}CoO_2 + xLiC_6 = LiCoO_2 + xC_6$. In this exercise, $x$ is 0.65. Thanks to the efforts of battery engineers and researchers, this fraction $x$ has increased with time. This means that the same amount of $LiCoO_2$ provides more capacity, or less amount of $LiCoO_2$ is sufficient to provide the same capacity. As $x$ increases over technology generations, the charge cutoff voltage increases, bringing both higher capacity and higher average voltage for the battery as shown in Table 2.2. This has been an important lever to increasing battery energy density over the years.

**Table 2.2**
An Example of Transition in Charge Cutoff Voltage and Nominal Voltage for a Li-ion Battery with $LiCoO_2$ Cathode and Graphite Anode

| | **Charge Cutoff Voltage** | **Nominal Voltage** |
|---|---|---|
| **Generation 1** | 4.2V | 3.7V |
| **Generation 2** | 4.35V | 3.80V |
| **Generation 3** | 4.4V | 3.85V |
| ⋮ | ⋮ | ⋮ |

## 2.4 SUMMARY

In this chapter, we learned the following:

- How to convert Gibbs free energy difference into battery voltage with the Nernst equation.
- How to calculate battery voltage with a standard potential table.
- How to calculate the required mass of cathode and anode to meet the battery life goal.

## 2.5 PROBLEMS

### Problem 2.1

You are planning to make a new battery with metal M and oxygen. In this battery, the anode is metal M and the cathode is oxygen. Metal M reacts with oxygen at the cathode and forms M oxide. The full reaction is $M + \frac{1}{2}O_2 = MO$ where M releases 2 electrons. Based on the reports from other researchers, you know that $\Delta G$ of this reaction is –609 kJ/mol. What is the theoretical battery voltage at the standard condition? Use 96485 C/mol for the Faraday constant.

### Answer 2.1

With the Nernst equation, $-609\times 10^3$ J/mol $= -2 \times 96485 \times E$. Solving this gives $E \approx 3.16$V.

Problem 2.2

Calculate the overall cell potential of the following reaction with Table 2.1.

$$Mg^{2+} + 2Li = Mg + 2Li^{+}$$

Answer 2.2

From Table 2.1, the following equations are derived.

$$Mg^{2+} + 2e^{-} = Mg \quad -2.36V$$

$$2Li = 2Li^{+} + 2e^{-} \quad +3.04V$$

Combining these equations leads to –2.36 + 3.06 = 0.68V.

Problem 2.3

A lithium metal rechargeable battery consists of the following half reactions.

Anode: $Li = Li^{+} + e^{-}$
Cathode: $Li_{1\text{-}x}CoO_2 + xLi^{+} + xe^{-} = LiCoO_2$

Answer the following questions when Li is 6.94 g/mol, $LiCoO_2$ is 97.87 g/mol, and the Faraday constant is 96485 C/mol.

1. To provide 1 Ah, how many grams of lithium are needed at least at anode before discharging?
2. When $x$ is 0.6, how many grams of $LiCoO_2$ are needed at least to provide 1 Ah?

Answer 2.3

1. 1 Ah is equivalent to 1 A × 1 h = 1 C/sec × 3600 sec = 3600C. To provide 1 Ah, 3600C of electrons are required. 3600C equals (3600/96485) mol electrons. The ratio of electrons to lithium is 1:1. Therefore, the required Li is (3600/96485) × 6.94 ≈ 0.259g.
2. When $x$ is 0.6, the ratio of electrons to $LiCoO_2$ is 0.6:1. Therefore, the required $LiCoO_2$ is (3600/96485) × 97.87 × (1/0.6) ≈ 6.086g.

On a side note, the full reaction is $Li_{1-x}CoO_2 + xLi = LiCoO_2$. This means that theoretically lithium metal is not needed when a lithium metal rechargeable battery is made. All lithium in the anode comes from the $LiCoO_2$ cathode during charge. The details can be found in Chapter 5.

## References

[1] Bratsch, S. G., "Standard Electrode Potentials and Temperature Coefficients in Water at 298.15K," *Journal of Physical and Chemical Reference Data,* Vol. 8, No. 1, 1989, pp. 3–11.

[2] Asenbauer, J., et al., "The Success Story of Graphite as a Lithium-ion Anode Material–Fundamentals, Remaining Challenges, and Recent Developments Including Silicon (Oxide) Composites," *Sustainable Energy Fuels,* Vol. 4, No. 11, 2020, pp. 5387–5416.

[3] Matsumura, N., "Cathode Technologies for Consumer Electronics," *Cathodes 2017,* California, USA, 2017, pp. 5–6.

[4] Aguiló-Aguayo, N., et al., "Water-Based Slurries for High-Energy $LiFePO_4$ Batteries Using Embroidered Current Collectors," *Scientific Reports,* Vol. 10, 2020, p. 5565.

[5] Schreiner, D., et al, "Comparative Evaluation of LMR-NCM and NCA Cathode Active Materials in Multilayer Lithium-Ion Pouch Cells: Part I. Production, Electrode Characterization, and Formation," *Journal of the Electrochemical Society,* Vol. 168, 2021, p. 030507.

[6] Li, J., et al, "Study of the Failure Mechanisms of $LiNi_{0.8}Mn_{0.1}Co_{0.1}O_2$ Cathode Material for Lithium Ion Batteries," *Journal of the Electrochemical Society,* Vol. 162, 2015, p. A1401.

[7] Placke, T., et al., "Lithium Ion, Lithium Metal, and Alternative Rechargeable Battery Technologies: The Odyssey for High Energy Density," *Journal of Solid State Electrochemistry,* Vol. 21, 2017, pp. 1939–1964.

[8] Nishi, Y., "Lithium Ion Secondary Batteries; Past 10 Years and the Future," *Journal of Power Sources,* Vol. 100, No. 1–2, 2001, pp. 101–106.

[9] Vadlamani, B., "An In-Situ Electrochemical Cell for Neutron Diffraction Studies of Phase Transitions in Small Volume Electrodes of Li-Ion Batteries," *Journal of the Electrochemical Society,* Vol. 161, 2014, pp. A1731–A1741.

[10] Gu, M., et al., "Nanoscale Silicon as Anode for Li-ion Batteries: The Fundamentals, Promises, and Challenges," *Nano Energy,* Vol. 17, 2015, pp. 366–383.

[11] Zhu, G., et al., "Dimethylacrylamide, a Novel Electrolyte Additive, Can Improve the Electrochemical Performances of Silicon Anodes in Lithium-ion Batteries," *RSC Advances,* Vol. 9, 2019, pp. 435–443.

[12] Cho, M., et al., "Anomalous Si-Based Composite Anode Design by Densification and Coating Strategies for Practical Applications in Li-ion Batteries," *Composites Part B: Engineering,* Vol. 215, 2021, p. 108799.

[13] Xu, W., et al., "Lithium Metal Anodes for Rechargeable Batteries," *Energy and Environmental Science,* Vol. 7, 2014, pp. 513–537.

[14] Zhou, L., et al., "Lithium Sulfide as Cathode Materials for Lithium-Ion Batteries: Advances and Challenges," *Journal of Chemistry,* Vol. 2020, Article ID 6904517.

[15] Bruce, P., et al., "Li-$O_2$ and Li-S Batteries with High Energy Storage," *Nature Materials,* Vol. 11, No. 1, 2012, pp. 19–29.

[16] Lide, D., *CRC Handbook of Chemistry and Physics, 86th Edition 2005–2006,* Boca Raton, FL: CRC Press, Taylor & Francis, 2005, pp. 4–70.

[17] Julien, C., "Lithium Iron Phosphate: Olivine Material for High Power Li-Ion Batteries," *Research & Development in Material Science,* Vol. 2, 2017.

# 3

# BATTERY IMPEDANCE AND ITS IMPACT ON BATTERY LIFE

## 3.1 INTRODUCTION

Design of systems, such as laptop PCs and electric vehicles, and development of their batteries, are always battles against impedance. For any system, battery life, which is defined as how long a battery will run on a full charge, is desired to be long. Impedance reduces battery life. For example, you may have an experience where a smartphone suddenly shuts down when you take a photo while building a snowman in winter. This is because of battery impedance. Performance of the system is also affected by battery impedance. For example, your smartphone response may be delayed when the battery comes close to empty. This chapter explains the details of battery impedance and its impact on battery life.

## 3.2 BATTERY IMPEDANCE

### 3.2.1 Ohm's Law and IR Drop

Before explaining battery impedance, let's review Ohm's law, which readers may already know. Ohm's law is a formula that explains the relationship between voltage, current, and resistance.

$$\text{Ohm's Law: } V = I \times R \tag{3.1}$$

where $V$ is voltage, $I$ is current, and $R$ is resistance.

Precisely speaking, battery impedance is different from resistance because it includes not only resistance but also capacitive reactance, which is another opposition to current flow. For convenience's sake, this Section 3.2.1 considers only resistance and treats resistance as impedance.

Figure 3.1 is an example of current flow from a battery to a system. This figure shows that the battery consists of an energy storage portion and a resistance portion, which is impedance. When the system uses energy from the battery, current goes from the battery to the system and returns to the battery. In the figure, the battery is at 3.8V and has 0.1 ohm internal impedance. When 1A is supplied from the battery, internal impedance takes 0.1V calculated as 1A × 0.1 ohm because of Ohm's law. This means that only 3.7V is available at the system because 0.1V out of 3.8V is used within the battery. Since the battery voltage drops by the amount of $I \times R$, such voltage drop is called IR drop.

To understand IR drop better, water is used as an analogy as shown in Figure 3.2. In this chart, voltage is height, current is water flow, and battery is water pump. First, water is pumped up to 3.8V by the energy storage portion of the battery. Then, water flows down through the battery internal impedance and system. 0.1 ohm of battery

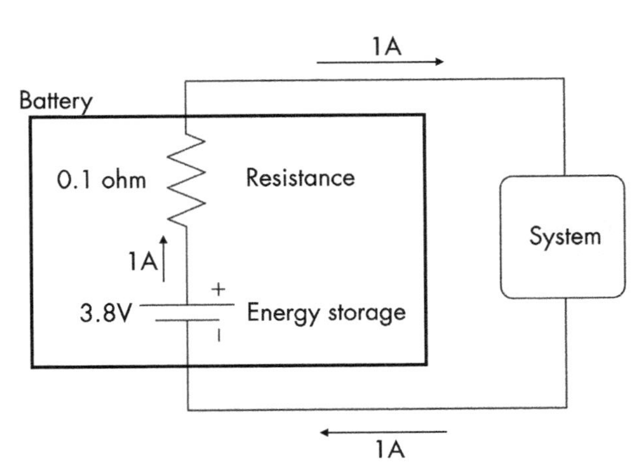


**Figure 3.1** Schematic illustration of current flow from a battery to a system.

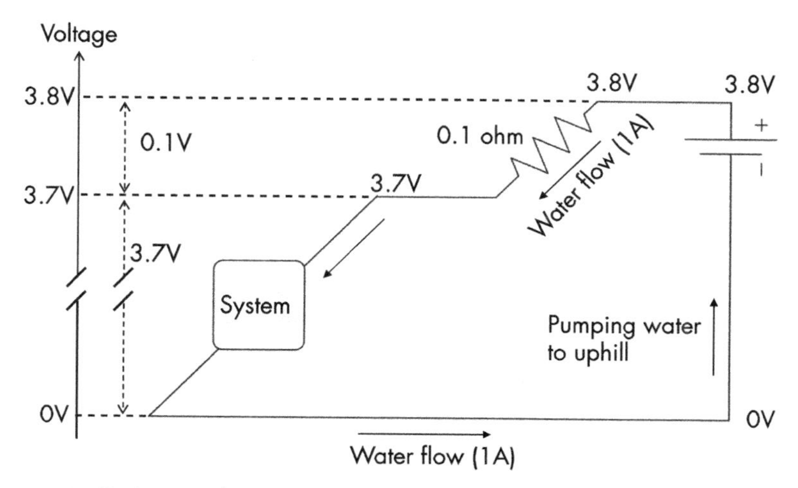


**Figure 3.2** IR drop explanation with water flow as an analogy.

impedance takes 0.1V in the case of 1A. Therefore, battery voltage is decreased to 3.7V. The system uses all 3.7V and voltage is decreased to 0V. In the next cycle, voltage is increased again to 3.8V by the battery and goes down.

Now let's look at how the battery voltage of a real Li-ion battery changes in Figure 3.3. In this figure, battery voltage at point A starts under no current at point A′. This means that the battery voltage is

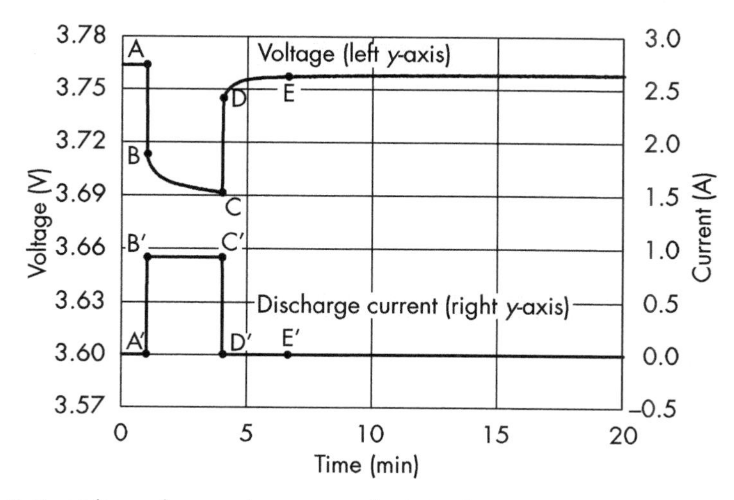


**Figure 3.3** Voltage drop and recovery of a Li-ion battery.

open-circuit voltage (OCV). When discharge starts at ~0.9A, which is 0.2C in this case, current increases to point B′ and battery voltage immediately drops to point B due to IR drop. After that, the battery voltage continues to decrease gently to point C while the current stays at the same from point B′ to point C′. When the current stops at point D′, the battery voltage immediately increases to point D because of recovery from IR drop. After that, battery voltage continues to recover gently to point E, and thereafter towards OCV for the next one hour, although current stays at zero. Why is the recovery of the battery voltage slow? If battery impedance consists of only resistance, such slow-voltage recovery after point D does not happen because response of resistance to voltage is immediate. The next section answers the question by explaining the details of battery impedance.

### 3.2.2 Equivalent Circuit Model

During charge and discharge of a Li-ion battery, lithium ions and electrons move between cathode and anode as explained in Chapter 1. There are several resistances in the ion and electron paths, for example, ion diffusion in cathode and anode materials, ion transfer through the electrolyte between cathode and anode, and electron transfer in cathode and anode. Not only resistances, but there are also capacitive reactances in the ion and electron paths, such as in cathode, anode, and those interfaces with the electrolyte [1–3].

To understand battery impedance, including both resistance and capacitive reactance, several electrical models have been proposed [1–4]. Such models are called equivalent circuit models. Figure 3.4 is an example of an equivalent circuit model.

This example model consists of a resistor, $R_0$, two parallel resistor-capacitor (RC) circuits, and an energy storage portion, $V_0$. The first parallel RC circuit consists of a resistor, $R_1$, and a capacitor, $C_1$,

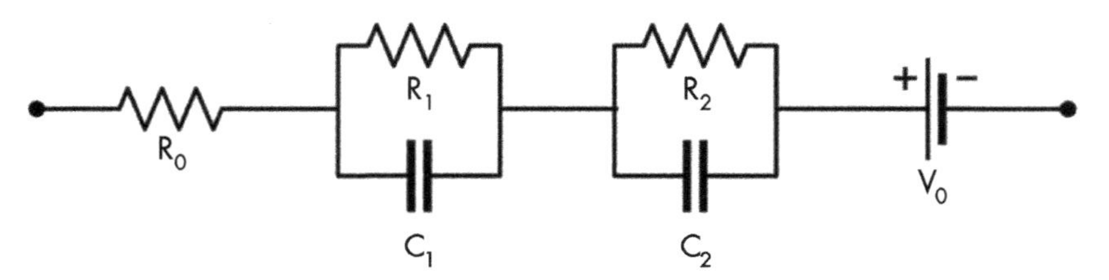


**Figure 3.4** An example of an equivalent circuit model.

connected in parallel. Similarly, the second parallel RC circuit consists of a resistor, $R_2$, and a capacitor, $C_2$, connected in parallel. While this model includes two parallel RC circuits, some models may include only one parallel RC circuit to simplify the model and reduce the computational resource to calculate the values of resistance and capacitance. Other models may include more than two parallel RC circuits. In the energy storage portion, $V_0$ is equal to OCV. In this book, a battery symbol is used for the energy storage portion to understand it intuitively, while some may use a different symbol, such as a voltage source, instead. $R_0$ typically represents resistance of ion transfer in electrolyte, and electron movement in metal conductors of cathode and anode, which are called current collectors. Capacitive reactance is caused by polarization of ions and electrons in cathode, anode, and their interfaces with electrolyte during charge and discharge, where there are also resistances. Therefore, parallel RC circuits represent those areas that include both resistance and capacitive reactance. Figure 3.5 is an example of a simplified equivalent circuit model where each circuit symbol shows the corresponding part in the schematic illustration of a Li-ion battery.

Often $R_1$ and $R_2$ are associated with charge transfer resistances, while $C_1$ and $C_2$ are associated with interfacial capacitances such as double-layer capacitances.

The equivalent circuit model well explains the slow voltage response to the current change observed in Figure 3.3. Before discharging starts at point A in Figure 3.3, the battery corresponds to Figure 3.5(a) where there is no charge in $C_1$ and $C_2$. The observed voltage at point A in Figure 3.3 is OCV, which is $V_0$ in Figure 3.5(a). When discharge starts at point B' in Figure 3.3, current I flows and resistance $R_0$ immediately takes the voltage, which is $I \times R_0$ as shown in Figure 3.5(b). At this point, capacitors $C_1$ and $C_2$ are at the beginning of charging and there are little capacitive reactances in two parallel RC circuits. Therefore, IR drop is only $I \times R_0$ and the observed voltage is $V_0 - I \times R_0$. This is the voltage at point B in Figure 3.3. During discharge, capacitors are charged in opposite polarity compared to $V_0$ as shown in Figure 3.5(c). This causes additional IR drops at RC circuits. If discharging continues, IR drops at the $R_1C_1$, and $R_2C_2$ circuits will increase to $I \times R_1$ and $I \times R_2$, respectively. As a result, the observed voltage will decrease to $V_0 - I \times R_0 - I \times R_1 - I \times R_2$. Note that the

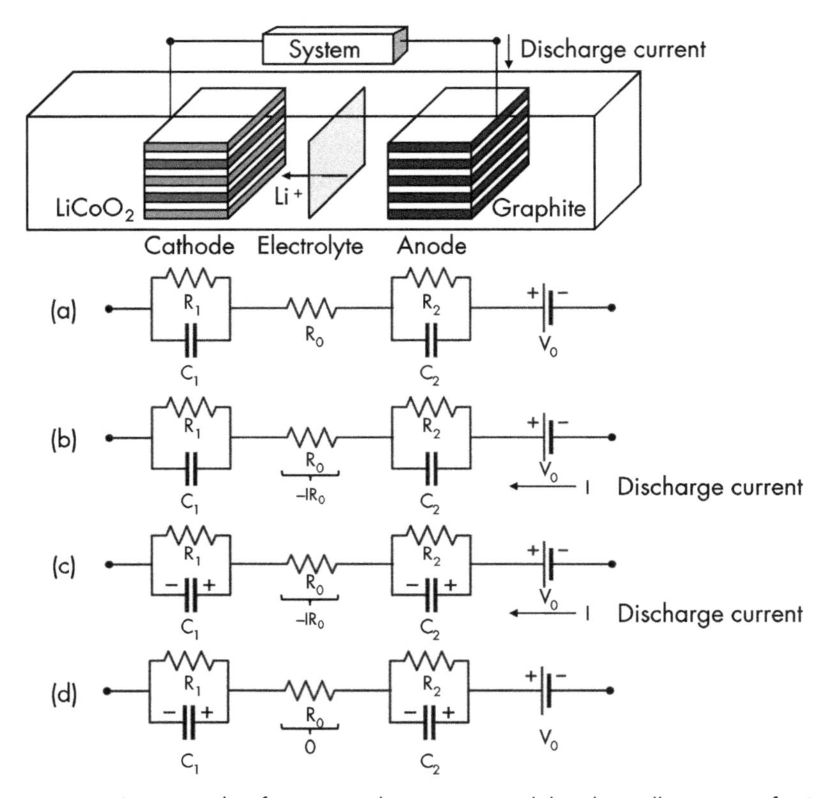


**Figure 3.5** An example of an equivalent circuit model with an illustration of a Li-ion battery: (a) at open circuit, (b) at initial discharge state, (c) during discharge, and (d) after discharge stops.

voltage drop to point C in Figure 3.3 also includes the voltage drop due to SOC change and the associated $V_0$ change. However, such SOC change causes only 0.0047V drop in this case, compared to 0.0725V drop from point A to C in Figure 3.3. This explains how largely battery impedance affects battery voltage drop.

When battery discharge stops at point D′ in Figure 3.3, the battery voltage sharply increases to point D. This is because one of the IR drops, $I \times R_0$, immediately becomes zero as shown in Figure 3.5(d) when discharge current stops. Still, the charges in $C_1$ and $C_2$ remain. These charges are slowly discharged within the parallel RC circuit because of the connected resistance. This is the reason why the voltage slowly recovers to point E and thereafter to OCV in Figure 3.3.

A battery management system needs to know OCV as early as possible. This is because OCV indicates remaining battery life, such as video play time of a smartphone, and driving range of an electric vehicle. While the estimation method is explained in Chapter 7, this section explains a practical model of voltage recovery.

### 3.2.2.1 *Equations for a Parallel RC Circuit*

In a parallel RC circuit, if there is no current input from the outside of the circuit, the voltage of the charged capacitor is decreased by the resistor in the following equation:

$$V_t = V_{ini} \exp\left[-t/(RC)\right] \tag{3.2}$$

where $V_t$ is the voltage at time $t$, $V_{ini}$ is the initial voltage of the capacitor, $R$ is resistance, and $C$ is capacitance of the parallel RC circuit.

With (3.2), the battery voltage after point D in Figure 3.3 can be expressed as follows:

$$V_{batt} = V_0 - V_{ini,1} \exp\left[-t/(R_1 C_1)\right] - V_{ini,2} \exp\left[-t/(R_2 C_2)\right] \tag{3.3}$$

where $V_{batt}$ is the observed battery voltage at time $t$, $V_0$ is OCV, $V_{ini,1}$ is the initial voltage of the parallel $R_1C_1$ circuit, $V_{ini,2}$ is the initial voltage of the parallel $R_2C_2$ circuit, $R_1$ and $R_2$ are resistance, and $C_1$ and $C_2$ are capacitance. All values in (3.3) are associated with Figure 3.5(d).

While the values of initial voltage, resistance, or capacitance of the parallel RC circuits are unknown, the coefficients in (3.3) can be obtained by fitting the equation to the data in Figure 3.3. Figure 3.6 shows the comparison of voltage recovery from IR drop between the measured data and the simulated data by the equivalent circuit model of (3.3).

In this figure, parameters of (3.3) were fit using only the first-two-minutes data from point D and did not change thereafter. In this case, the measured OCV was 3.759V. To measure OCV, it took ten minutes. It may take hours in some cases. This simulation predicted that OCV is 3.757V with only two minutes of data. This is close enough for some use cases. The accuracy may be different when the conditions are different, such as after longer charging or discharging, at different temperatures, and at different SOC. There are many other models that

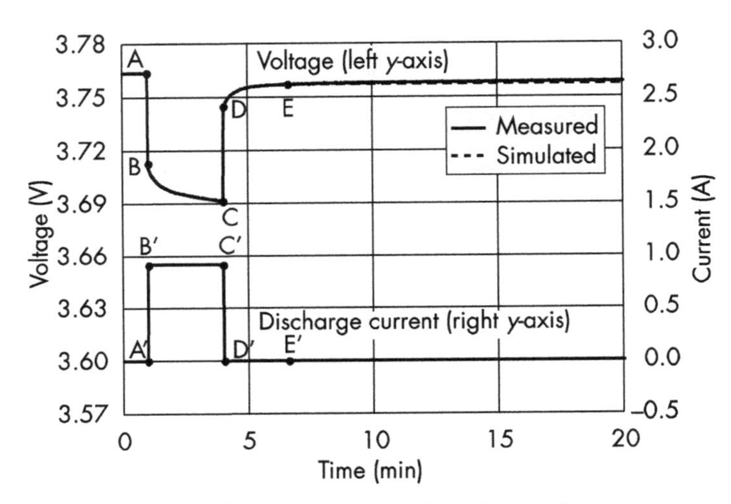


**Figure 3.6** Comparison of voltage recovery from IR drop between measured data and the simulated data by an equivalent circuit model.

may make the results better. The key takeaway in this section is that a battery impedance is complicated but can be expressed by equivalent circuit models. The models not just explain the observed battery data, but also predict the future battery characteristics such as OCV.

### 3.2.3 Impedance Measurement Method by Electrochemical Impedance Spectroscopy

After learning equivalent circuit models, you may be interested to know if there is a method to measure resistance and capacitance in the models. The answer is electrochemical impedance spectroscopy (EIS). EIS is a technique that applies a small alternating current (AC) signal to a device, which is a battery in this book, and measures its response. Since physical processes in a battery span a wide range of time scales (electronic movements, charge transfer at surfaces, diffusion over longer time scales, etc.), EIS distinguishes them by sweeping the applied frequency. As the details of EIS are mathematically complex, this section focuses on basic theory necessary for a battery and its application to fitting of equivalent circuit models.

To apply an AC to a battery, a frequency response analyzer (FRA) is used. FRA applies a small AC wave, such as 10 mV, over wide frequency ranges, such as 100 kHz to 10 mHz.

#### 3.2.3.1 Theory

When an AC is applied at angular frequency $\omega$, and voltage amplitude $E_0$ for example, the applied voltage $E_t$ is expressed with a complex function as:

$$E_t = E_0 \exp(j\omega t) \tag{3.4}$$

where $j$ is the imaginary unit $\sqrt{-1}$ and $t$ is time. $\omega$ can be replaced with $2\pi f$ where $f$ is the ordinary frequency in hertz.

You may be surprised to see $j$, the imaginary unit, in the practical engineering equation. This is to treat AC and impedance mathematically easily. If you don't feel comfortable, you may skip the theory section and move to the application section. On a side note, $i$ is typically used for the imaginary unit. However, $i$ is also used for current. To avoid confusion, $j$ is used for the imaginary unit.

Because of resistances and reactances in a battery, the AC phase is shifted by $\theta$ and the measured current response $I_t$ is expressed as:

$$I_t = I_0 \exp[j(\omega t - \theta)] \tag{3.5}$$

where $I_0$ is the current amplitude.

And the impedance is expressed as follows:

$$\begin{aligned} Z &= E_t / I_t \\ &= Z_0 \exp(j\theta) \end{aligned} \tag{3.6}$$

where $Z_0$ is the magnitude of impedance.

With Euler's formula, (3.6) leads to the following:

$$Z = Z_0 (\cos\theta + j\sin\theta) \tag{3.7}$$

Depending on the input frequency, AC phase shift $\theta$ changes, resulting in impedance change.

When the magnitude of the impedance is plotted against the input frequency, the graph is called a Bode plot. Figure 3.7 is an example of a Bode plot with a Li-ion battery.

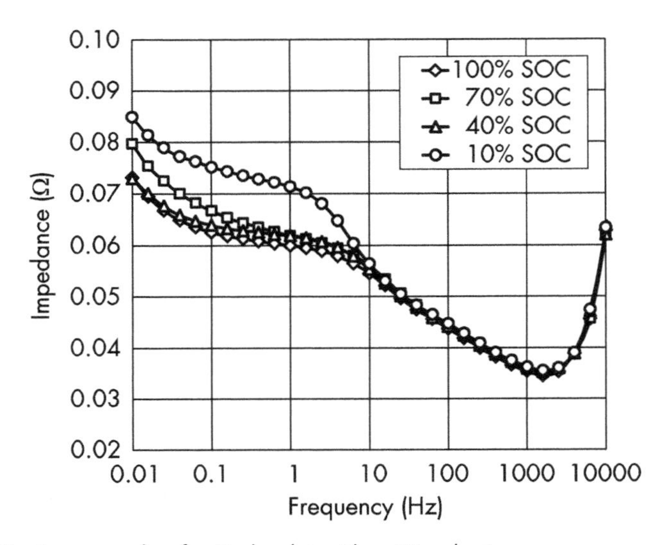


**Figure 3.7** An example of a Bode plot with a Li-ion battery.

As the figure shows, the magnitude of impedance depends on frequencies. This is because impedance is caused by transfer and polarization of ions and electrons, and response behavior of ions and electrons depends on the applied frequency. The response behavior is also affected by SOC, resulting in the impedance change.

Impedance $Z$ can also be expressed as:

$$Z = R + jX \tag{3.8}$$

where $R$ and $X$ are the real part and the imaginary part of impedance, respectively. Note that the names of real and imaginary are from a mathematical standpoint and not literally "real" or "imaginary." Both exist in the battery.

By comparing (3.7) and (3.8), you will notice that

$$R = Z_0 \cos\theta \tag{3.9}$$

$$X = Z_0 \sin\theta \tag{3.10}$$

With EIS, $Z_0$ and $\theta$ are measurable. Therefore, $R$ and $X$ can be determined.

*3.2.3.2 Equivalent Circuit Model with Complex Function*

With complex function, the impedance of equivalent circuit elements is expressed as:

Resistor: $R_1$, where $R_1$ is resistance

Capacitor: $1/(j\omega C_1)$, where $C_1$ is capacitance

With these, impedance of a parallel RC circuit, $Z$(parallel), is calculated as:

$$\begin{aligned} Z(parallel) &= \{R_1 \times [1/(j\omega C_1)]\}/[R_1 + 1/(j\omega C_1)] \\ &= R_1/(1 + j\omega R_1 C_1) \\ &= R_1/(1 + \omega^2 R_1^2 C_1^2) + j(-\omega R_1^2 C_1)/(1 + \omega^2 R_1^2 C_1^2) \end{aligned} \tag{3.11}$$

This means that impedance of a parallel RC circuit can be expressed with the real and imaginary parts as shown in (3.8).

Then, let's think about a simple equivalent circuit model shown in Figure 3.8, often referred to as a simplified Randles cell.

This model includes one resistor and one parallel RC circuit. The total impedance $Z_{total}$ can be expressed as:

$$Z_{total} = R_0 + Z(parallel) \tag{3.12}$$

With (3.11) and (3.12),

$$Z_{total} = [R_0 + R_1/(1 + \omega^2 R_1^2 C_1^2)] + j(-\omega R_1^2 C_1)/(1 + \omega^2 R_1^2 C_1^2) \tag{3.13}$$

Equation (3.13) consists of the real and imaginary parts, which is similar to (3.8). With EIS, R and X in (3.8) are measurable.

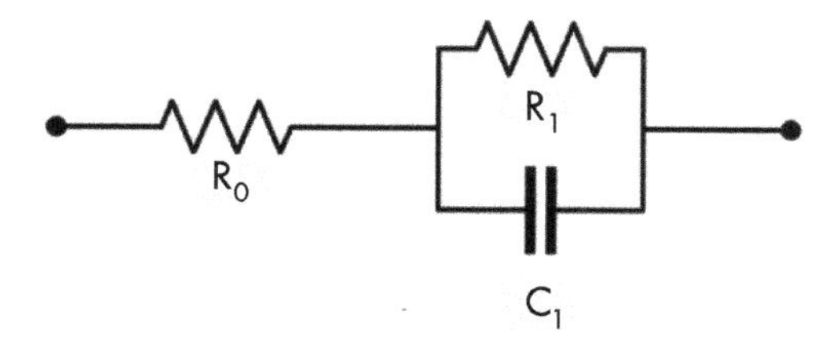


**Figure 3.8** A simple equivalent circuit model with one resistor and one parallel RC circuit.

Figure 3.9 shows R and X plotted in horizontal axis and vertical axis, respectively, at various frequencies.

In Figure 3.9, the horizontal axis is the real part of the impedance, Z′. The vertical axis is the imaginary part of the impedance with negative sign, –Z″. This means that Z′ is R and –Z″ is –X. The sign of X is reversed for easier analysis. This plot is called the Nyquist plot.

*3.2.3.3 Calculation of Resistance and Capacitance from the Nyquist Plot*

From (3.8) and (3.13), the following is derived:

$$R = R_0 + R_1/\left(1+\omega^2 R_1^2 C_1^2\right) \tag{3.14}$$

$$X = \left(-\omega R_1^2 C_1\right)/\left(1+\omega^2 R_1^2 C_1^2\right) \tag{3.15}$$

By combining (3.14) and (3.15) to remove $\omega$, the following equation is given:

$$\left[R-\left(R_0+R_1/2\right)\right]^2 + X^2 = \left(R_1/2\right)^2 \tag{3.16}$$

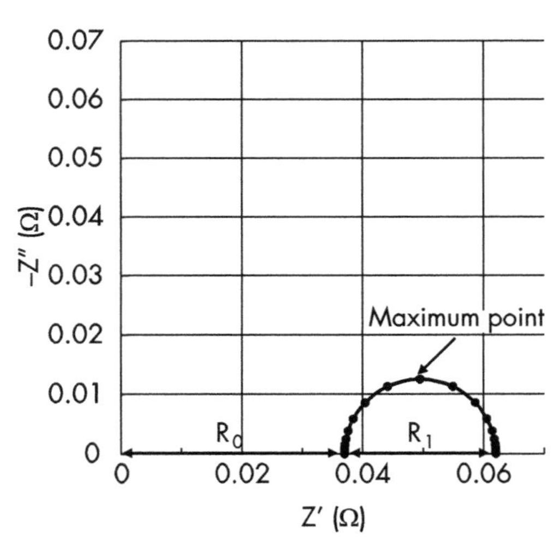


**Figure 3.9** The real and imaginary parts of impedances for a circuit with one resistor and one parallel RC circuit.

This means that the coordinates of the center in the semicircle of Figure 3.9 are $(R_0 + R_1/2, 0)$ and the radius is $R_1/2$. Therefore, $R_0$ corresponds to the horizontal value of the left side in the semicircle, and $R_1$ corresponds to the diameter of the semicircle. Also, at the maximum point of the semicircle, the coordinates are $(R_0 + R_1/2, R_1/2)$. Inserting this to either (3.14) or (3.15) gives $\omega = 1/(R_1C_1)$. As $\omega$ equals $2\pi f$ where f is the frequency at the maximum point and $R_1$ is already known as the diameter of the semicircle, $C_1$ can also be derived. In this case, $R_0$ is 0.037 ohm, $R_1$ is 0.025 ohm, the frequency f at the maximum point is 15.8 Hz, and $C_1$ is calculated as 0.40F.

*3.2.3.4 Application to the Nyquist Plot of the Real Li-ion Battery*

Figure 3.10 shows a Nyquist plot of a real Li-ion battery cell with 3 Ah. Point A is measured at high frequency. As the frequency decreases, the measured data goes to point B and a straight line with a slope of 45 degrees. The straight line is due to Warburg impedance, which is caused by the polarization in the lithium diffusion reactions in electrodes [5]. While this is useful information to analyze the diffusion coefficient of lithium in cathode and anode materials, this section focuses on the data from point A to point B. Unlike Figure 3.9, the shape from point A to point B does not look like a semicircle. This is because the region consists of the combination of multiple semicircles. One semicircle corresponds to one parallel RC circuit as

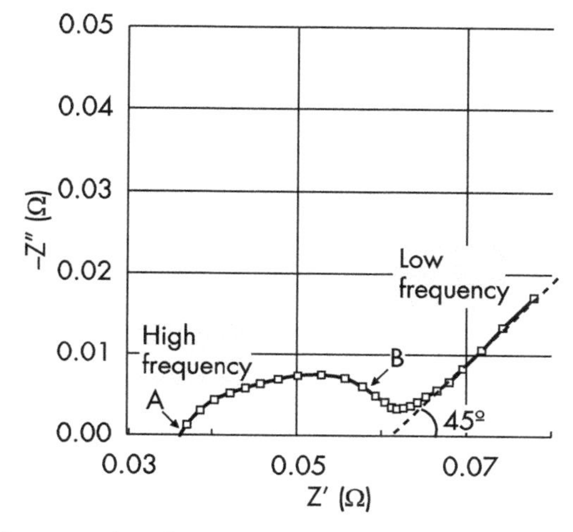


**Figure 3.10** A Nyquist plot of a 3-Ah Li-ion battery cell.

explained earlier. When an equivalent circuit model has one resistor and two parallel RC circuits as shown in Figure 3.4, the shape from point A to point B in Figure 3.10 consists of the combination of two semicircles. In that case, by adding another parallel RC circuit with resistance $R_2$ and capacitance $C_2$ to (3.13), the total impedance, $Z_{total}$, can be expressed as

$$\begin{aligned} Z_{total} &= \left[R_0 + R_1/\left(1+\omega^2 R_1^2 C_1^2\right) + R_2/\left(1+\omega^2 R_2^2 C_2^2\right)\right] \\ &+ j\left[\left(-\omega R_1^2 C_1\right)/\left(1+\omega^2 R_1^2 C_1^2\right) + \left(-\omega R_2^2 C_2\right)/\left(1+\omega^2 R_2^2 C_2^2\right)\right] \end{aligned} \tag{3.17}$$

As explained in Section 3.2.2, $R_0$ is associated with resistance of ion transfer in electrolyte, and electron movement in metal conductors of cathode and anode. $R_1$ and $R_2$ are associated with charge transfer resistances, while $C_1$ and $C_2$ are associated with interfacial capacitances.

When $R_0$ is 0.037 ohm, $R_1$ is 0.008 ohm, $C_1$ is 1F, $R_2$ is 0.015 ohm, and $C_2$ is 7F, the simulated data is shown in Figure 3.11, together with the measured data.

As shown in Figure 3.11, the equivalent circuit model well explains the behavior of a battery impedance. When more parallel RC

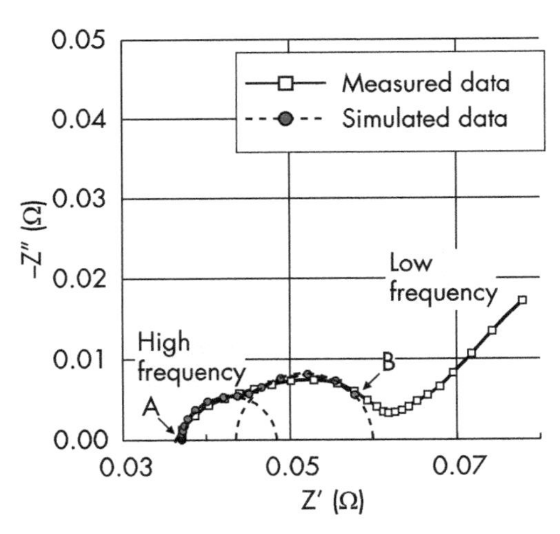


**Figure 3.11** A Nyquist plot of measured and simulated impedances for a 3-Ah Li-ion battery cell.

circuits are added in the model or the model is modified, even better fitting is possible.

In this section, we learned how to measure resistance and capacitance values in the equivalent circuit model with EIS. The theory and method are applicable not only to the simple circuit model used in this section, but also other complex models. For example, you may find a model that also includes an inductor that is expressed as $j\omega L$, where L is inductance. A more complex circuit model makes analysis more difficult. Some software detects the semicircles in the Nyquist pilot and calculates the characteristics of the components in your equivalent circuit model. Using such software is also an option.

### 3.2.4 AC Impedance and DC Impedance

Thorough analysis on battery impedance with EIS is informative. However, it requires an expensive FRA. It also takes time and human resources. In the industry, it is required to know the impedance instantly for several use cases, such as quality check in a battery manufacturing process, and battery management system. How does the industry complete such tasks? This section introduces two impedances, AC impedance and direct current (DC) impedance. These are also called AC resistance (ACR) and DC resistance (DCR), respectively. Both are widely used in the industry but require accurate knowledge to avoid misunderstanding.

#### *3.2.4.1 AC Impedance*

AC impedance is measured by applying AC wave at a specific frequency to a battery at specific SOC, whereas EIS uses a wide frequency range. The frequency to measure AC impedance is typically 1 kHz and a battery SOC is 50%. A battery cell spec shows an AC impedance limit, such as maximun 40 mohm. AC impedance can be measured instantly with an affordable tool for a quality check. For example, at the last stage of the battery manufacturing process, battery AC impedance is checked as a part of outgoing quality check and is confirmed not to exceed the limit written in the spec. In the spec, the measurement conditions need to be described such as AC 1 kHz and 50% SOC. This is because impedance depends on the frequency and SOC. Impedance also depends on temperature. Therefore, measurements need to be performed in a room where the temperature is controlled.

#### *3.2.4.2 DC Impedance*

DC impedance is the impedance measured after steady current for a certain period.

Figure 3.12 is an example of DC impedance measurement. At point A in Figure 3.12, the battery is at open circuit for a while and battery voltage is OCV. First, the battery is discharged at constant current, $I_b$, for a while to point B′. At the end of the discharging, the battery voltage is $V_b$ at point B. Then, discharging is stopped. After a certain period, the battery voltage is measured again as $V_c$ at point C. If the waiting period is long enough, $V_c$ is OCV. However, it may take a long time, such as hours. Therefore, different criteria may be used depending on the purpose of the impedance usage, such as measuring $V_c$ when voltage change for 10 minutes is within 2 mV. DC impedance is calculated as:

$$(\text{DC impedance}) = dV/dI = (V_c - V_b)/(I_b - 0) = (V_c - V_b)/I_b \quad (3.18)$$

For example, when $V_b$ is 3.69V, $V_c$ is 3.76V, and $I_b$ is 0.920A for 3 minutes, DC impedance is (3.76 – 3.69)/0.920 ≈ 0.0761 ohm. Note that the voltage of point A is not used for DC impedance calculation because SOC at point A is different from the SOC at point B or C.

DC impedance is useful when a system with a battery, such as a smartphone, uses DC from the battery, and IR drop under a certain

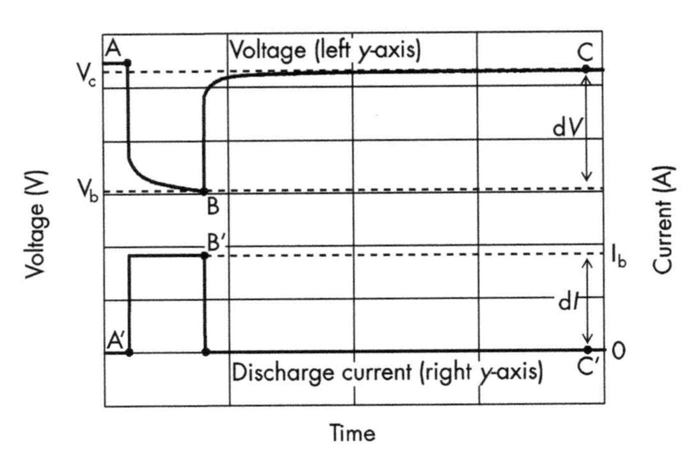


**Figure 3.12** An example of DC impedance measurement.

condition needs to be calculated easily to avoid hitting the system shutdown voltage. However, DC impedance depends on measurement conditions. For example, when a battery equivalent circuit model includes a parallel RC circuit as shown in Figure 3.4, its capacitive reactance of the battery impedance changes by duration and magnitude of current. DC impedance also changes by SOC, temperature, battery degradation status, initial battery status (which may be different from open circuit, and so forth.)

Figure 3.13 is an example of the difference between AC impedance at 1 kHz, DC impedance after 0.2C for 28 minutes, and DC impedance after 1C pulse for 1 msec.

As shown in the figure, there is a difference between AC impedance and DC impedances because of the difference in measurement methods. Furthermore, for the DC impedances, there is a difference depending on the discharging conditions. DC impedance after 0.2C for 28 minutes shows the highest impedance. This is because capacitive reactances of two parallel RC circuits in Figure 3.4 are higher after long discharging than the other two impedances in Figure 3.13.

When you calculate IR drop with DC impedance, it is important to confirm that the measurement conditions align with your use cases.

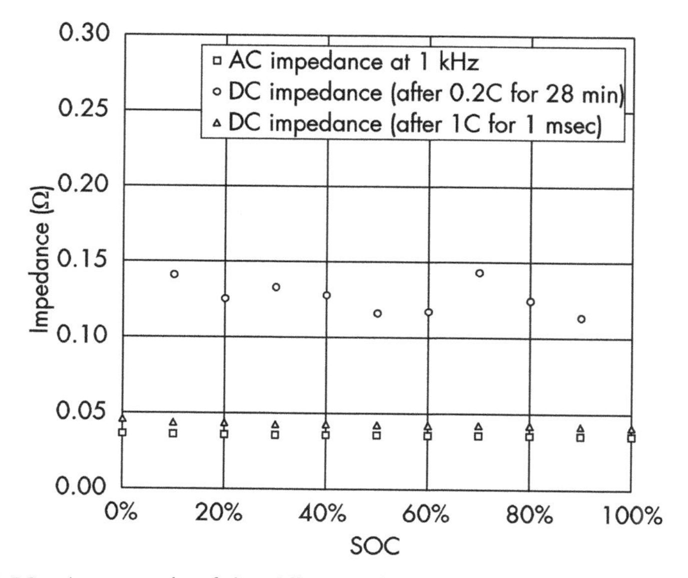


**Figure 3.13** An example of the difference between AC impedance at 1 kHz, DC impedance after 0.2C for 28 minutes, and DC impedance after 1C for 1 msec.

## 3.3 BATTERY DISCHARGING CHARACTERISTICS

### 3.3.1 Battery Discharging under Various Current Rates

We already learned that battery discharging causes IR drop and the observed voltage is lower than the actual battery voltage due to impedance. Figure 3.14 shows discharging curves of a Li-ion battery at various C-rates.

In Figure 3.14, at no current, battery voltage is OCV. This voltage decreases as discharged capacity increases. This is because, during discharge, $LiC_6$ anode changes towards $C_6$ and $Li_{1-x}CoO_2$ changes towards $LiCoO_2$, resulting in the change in the Gibbs free energy of cathode and anode. When discharging current increases from 0.2C to 1.0C, a larger IR drop is observed. This is because the IR drop is calculated as $I \times R$ where $I$ is discharging current, and $R$ is battery impedance. When $I$ increases, $I \times R$ increases. In this figure, the battery discharge cutoff voltage is 3.0V and the system shutdown voltage is 3.3V. When the battery voltage hits the system shutdown voltage during discharge, the system shuts down before hitting the battery discharge cutoff voltage. This means that higher discharging current hits the system shutdown voltage earlier due to a larger IR drop, resulting in less discharged capacity.

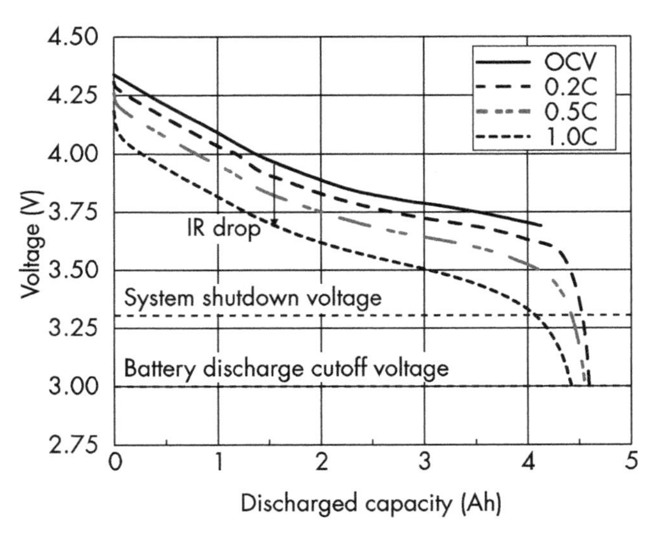


**Figure 3.14** An example of Li-ion battery discharging rate characteristics.

***Exercise***

Your smartphone response is suddenly delayed when the battery charge level is low. Explain the reason.

***Answer***

At the low charge level, battery voltage is close to system shutdown voltage. To avoid hitting the system shutdown voltage, one option is to reduce IR drop by limiting current. If this is the case, the limited current gives limited power to the processor in the smartphone, resulting in delayed response.

* * *

### 3.3.2 Battery Discharging at Various Temperatures

Battery capacity also depends on temperature. Figure 3.15 shows discharging curves of a Li-ion battery at various temperatures.

As shown in the figure, the discharging curves at lower temperatures show lower voltage. This is because, at low temperature, ionic conductivity becomes lower, impedance increases, and larger IR drop happens. When the temperature is even lower (e.g., –20°C in the

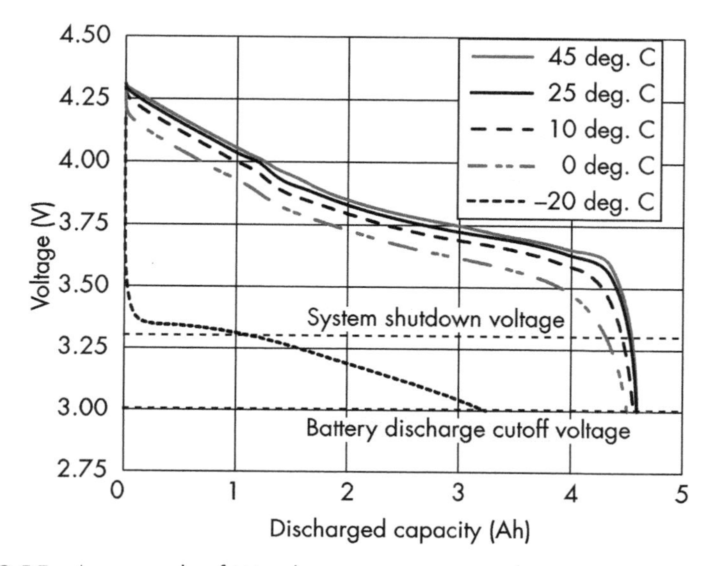


**Figure 3.15** An example of Li-ion battery temperature characteristics.

figure), the impedance increases significantly because the electrolyte liquidity is decreased. When the battery voltage hits the system shutdown voltage during discharge, the system shuts down before hitting battery discharge cutoff voltage. As a result, less capacity is usable at lower temperature.

> ***Exercise***
> While building a snowman in winter, you take your smartphone out of your pocket. The screen shows that battery SOC is 10%. However, it suddenly shuts down when you take a photo with the smartphone. Explain the reason.
>
> ***Answer***
> A possible reason is that, when the smartphone is taken out of the pocket, battery temperature is decreased due to a cold environment. Also, taking a photo requires higher current. These circumstances may result in larger IR drop and hit the system shutdown voltage earlier.

* * *

### 3.3.3 Impedance Dependency on Cycles

Battery impedance increases after charging and discharging cycles due to side reactions and growth of resistive layers in the cell. Figure 3.16 shows how battery impedance changes after charging and discharging cycles.

Figure 3.16(a) is AC impedance at 1 kHz for a 32-mAh Li-ion battery cell. Open circles are the initial impedances and solid circles are after 475 cycles of 0.5C CC-CV charge and 0.5C discharge. Figure 3.16(b) is DC impedance for the same 32-mAh cell. This figure shows two kinds of DC impedances: one measured after 0.2C discharging for 28 minutes, and the other one measured after 1C pulse discharging for 1 msec. For both DC impedances, initial data and after 475 cycles data is shown. In both Figures 3.16(a, b), battery impedance increases after cycles. This means that, after cycles, battery IR drop increases and battery voltage hits system shutdown voltage or discharge cutoff voltage earlier. This results in shorter battery life or worse system performance.

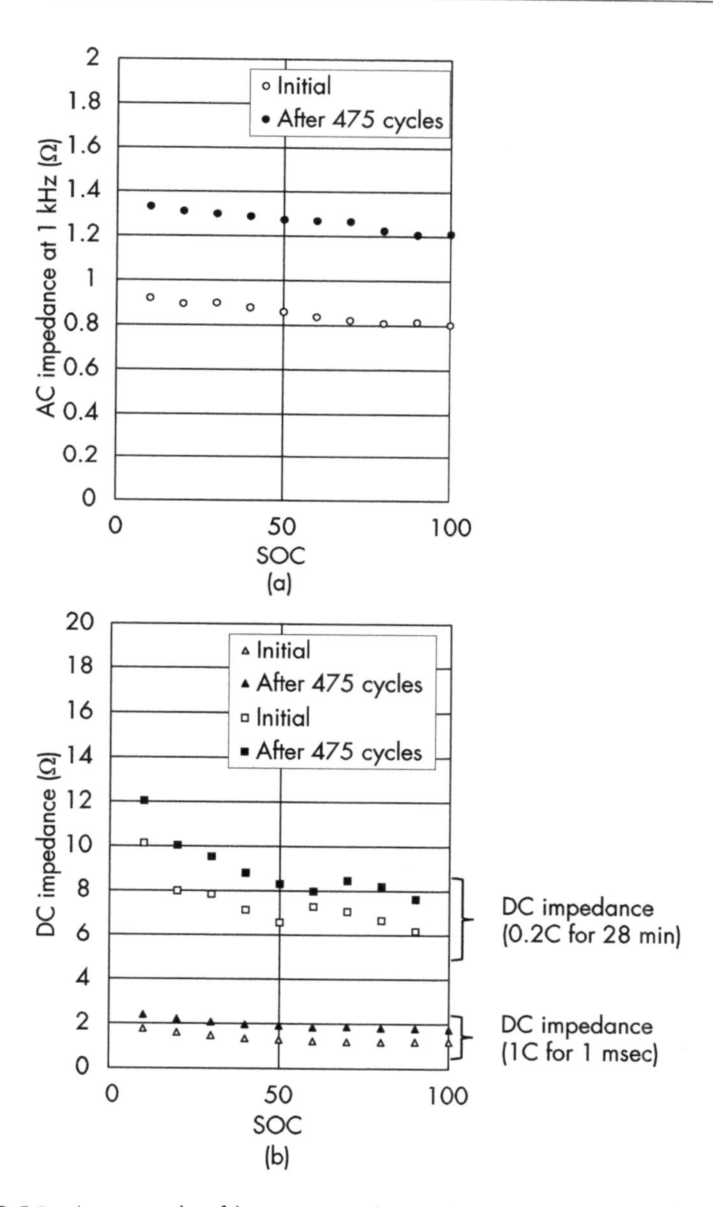


**Figure 3.16** An example of battery impedance changes over 475 cycles of 0.5C CC-CV charge and 0.5C discharge: (a) AC impedance at 1 kHz, and (b) DC impedance after 0.2C for 28 minutes or 1C for 1 msec.

Battery impedance also increases with storage time. The higher battery SOC during storage results in the higher degradation rate.

When a system with a Li-ion battery such as a smartphone and a smartwatch is manufactured and shipped from the factory, a low-battery SOC gives less impedance growth and less permanent capacity loss. However, a high SOC may be desired to meet out-of-box expectations by the users. For example, the users would like to power on and enjoy the new system when the box is first opened. This is usually a compromise reached between the battery engineer and the product owner who is responsible for determining overall user experience.

## 3.4 USABLE BATTERY CAPACITY

A battery spec or datasheet shows battery capacity, such as 0.41 Ah. What does this mean? It means that the capacity is measured under a standard test condition. The condition is, for example, 0.2C discharge to 3.0V cutoff voltage at 25°C. However, in a real system, discharging current may be higher, system shutdown voltage may be higher than battery discharge cutoff voltage, and the system with the battery may be used at low temperatures. These are all possible reasons for less usable capacity due to larger IR drop or less room to the discharging end voltage. Figure 3.17 shows an example of battery usage in a system.

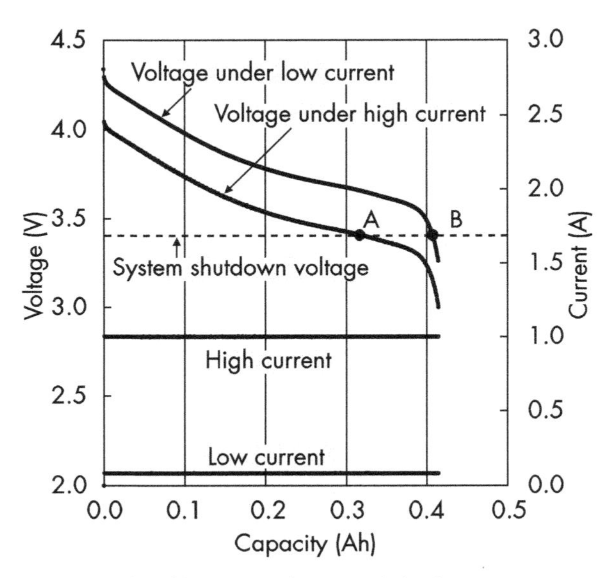


**Figure 3.17** An example of battery voltage and discharging current in a system.

In most of the systems, battery discharging current is not constant. It is sometimes high and sometimes low. In Figure 3.17, when discharging current is low, battery voltage is relatively high, and when discharging current is high, battery voltage is lower due to impedance. When the battery is under high current and hits system shutdown voltage at point A, the system shuts down and the usable battery capacity is 0.32 Ah, whereas the capacity in the spec is 0.41 Ah. If the system limits performance and reduces discharging current, such as low power mode, usable battery capacity may be extended to point B, but it is a trade-off against the performance. If battery impedance can be lowered by cell design or structure, more usable capacity is possible.

This section explained how IR drop decreases usable capacity in Ah. This is also the case for usable energy in Wh. When impedance is high, usable energy is less mainly due to two reasons. The first reason is due to hitting the system shutdown voltage earlier. In this case, part of energy is left unused, which is the similar reason to the decreased usable capacity. The second reason is the decreased battery voltage due to IR drop. During discharge, the lost energy caused by the voltage drop is dissipated as heat. The total lost energy due to voltage drop is calculated as:

$$(Lost\,energy) = \int_0^a IR\,dAh \tag{3.19}$$

where $a$ is the usable capacity, $I$ is current, $R$ is the battery impedance, and Ah is the discharged capacity.

This is equivalent to the shaded area in Figure 3.18 in case of 1C discharging.

In this case, the usable energy in case of 1C discharging, $E$, is expressed by the following equation and is shown as the dotted area in Figure 3.18:

$$E = \int_0^a V\,dAh \tag{3.20}$$

where $a$ is the usable capacity, $V$ is the battery voltage, and Ah is the discharged capacity.

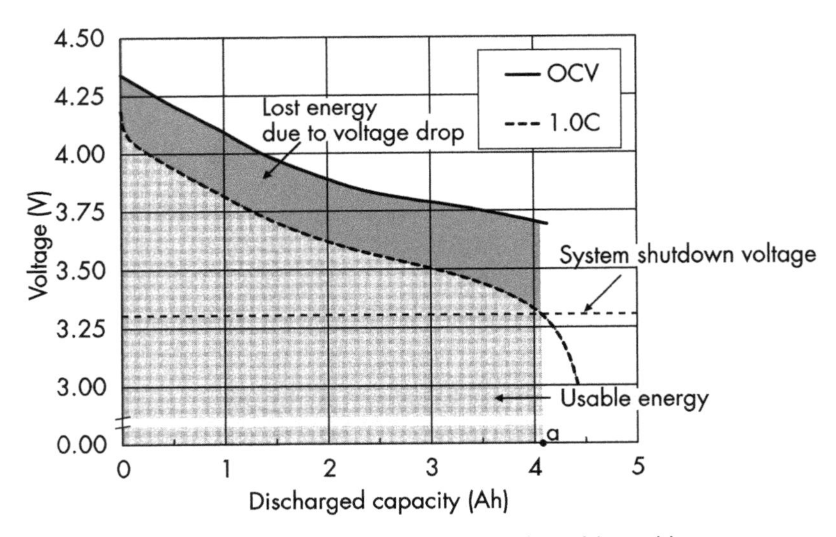


**Figure 3.18** A Li-ion battery discharging curve with usable and lost energy.

## 3.5 SUMMARY

In this chapter, we learned the following:

- Battery impedance causes IR drop during discharge, resulting in less battery life.
- IR drop is affected by discharging current, temperature, degradation status, discharging conditions, SOC, and so forth.
- Battery impedance can be expressed via equivalent circuit models.

The details of battery degradation over cycles and the methods to extend its longevity are explained in Chapter 10.

## 3.6 PROBLEMS

### Problem 3.1

There is a battery that shows 3.8V as OCV and has 0.05 ohm impedance. When 2A is discharged from the battery, what is the observed voltage at the battery?

Answer 3.1

3.8V – 2A × 0.05 ohm = 3.7V.

Problem 3.2

A battery is discharged at 2.5A for a while. After that, discharge stopped, and the battery voltage recovered from 3.75V to 4.00V. In this case, what is the DC impedance of the battery?

Answer 3.2

(4.00V – 3.75V) / 2.5A = 0.1 ohm

Problem 3.3

A system engineer is specifying maximum current that a system can draw from the attached battery. The engineer knows that the battery AC impedance in the spec shows 0.1 ohm. The system shutdown voltage is 3.3V. The engineer planned to intermittently use long 2A in the system and estimated that the system will shut down when battery OCV reaches around 3.5V because 3.5V – 2A × 0.1 ohm = 3.3V. However, the system shutdown at higher battery OCV than 3.5V. What was a possible reason?

Answer 3.3

There are several answers.

Example 1: In this case, the impedance used for the calculation needs to be DC impedance, which is different from AC impedance. As DC impedance after long discharging is higher than AC impedance, larger IR drop happens. This led to the system shutdown at higher battery OCV than expected.

Example 2: Battery temperature was lower than the spec and impedance was higher.

Example 3: Battery was degraded, and impedance was higher.

Note that in the system design, not only battery impedance but also other impedances in the current path need to be considered.

Problem 3.4

An engineer assumed that the equivalent circuit model of the battery consists of one resistor and two parallel RC circuits as shown in

Figure 3.4. When discharging of the battery stops and the battery voltage recovers to OCV, the model can be expressed as:

$$V_{batt} = V_0 - V_{ini,1}\exp\left[-t/(R_1C_1)\right] - V_{ini,2}\exp\left[-t/(R_2C_2)\right]$$

where $V_{batt}$ is the observed battery voltage at time t, $V_0$ is OCV, $V_{ini,1}$ is the initial voltage of the parallel $R_1C_1$ circuit, $V_{ini,2}$ is the initial voltage of the parallel $R_2C_2$ circuit, $R_1$ and $R_2$ are resistance, and $C_1$ and $C_2$ are capacitance of the $R_1C_1$ and $R_2C_2$ circuits, respectively. When the engineer performed the fitting of the model to the measured voltage data during voltage recovery after discharging stopped, the following equation was obtained:

$$V_{batt} = 3.758 - 1.026\times10^{-2}\exp(-t/0.3663) - 2.907\times10^{-3}\exp(-t/2.996)$$

What is the estimated OCV?

Answer 3.4

By comparing the model equation with the obtained equation after fitting, the OCV is estimated as 3.758V.

## References

[1] Osaka, T., et al., "Proposal of Novel Equivalent Circuit for Electrochemical Impedance Analysis of Commercially Available Lithium Ion Battery," *Journal of Power Sources,* Vol. 205, 2012, pp. 483–486.

[2] Momma, T., et al., "AC Impedance Analysis of Lithium Ion Battery Under Temperature Control," *Journal of Power Sources,* Vol. 216, 2012, pp. 304–307.

[3] Krewer, U., et al., "Review—Dynamic Models of Li-Ion Batteries for Diagnosis and Operation: A Review and Perspective," *Journal of the Electrochemical Society,* Vol. 165, No. 16, 2018, pp. A3656–A3673.

[4] Chen, M., et al., "Accurate Electrical Battery Model Capable of Predicting Runtime and I–V Performance," *IEEE Transactions on Energy Conversion,* Vol. 21, No. 2, 2006, pp. 504–511.

[5] Choi, W., et al., "Modeling and Applications of Electrochemical Impedance Spectroscopy (EIS) for Lithium-ion Batteries," *Journal of Electrochemical Science and Technology,* Vol. 11, No. 1, 2020, pp. 1–13.

# 4

# BATTERY CHARGING AND IMPEDANCE IMPACT

## 4.1 INTRODUCTION

Battery charging speed is an important feature of products, such as smartphones, laptop PCs, and electric vehicles. Faster charging gives better user experience. To enable fast charging, is it sufficient to increase charging current? The answer is no. It is important to consider impedance, safety, and many other factors. This chapter explains several battery charging methods and key design aspects, mainly for Li-ion batteries.

## 4.2 LI-ION BATTERY CHARGING

### 4.2.1 Constant Current-Constant Voltage Charging

For typical Li-ion batteries, CC-CV charging is used as explained in Section 1.5. CC and CV are the abbreviations for constant current and constant voltage, respectively. Figure 4.1 shows an example of CC-CV charging.

When SOC, that is battery charge level is low, CC-CV charging starts with CC charging, which is constant current. During charge, battery voltage and SOC increase. When the battery voltage reaches the charge cutoff voltage, charging mode shifts to CV, which is constant voltage. Charge cutoff voltage and the voltage at CV are the same.

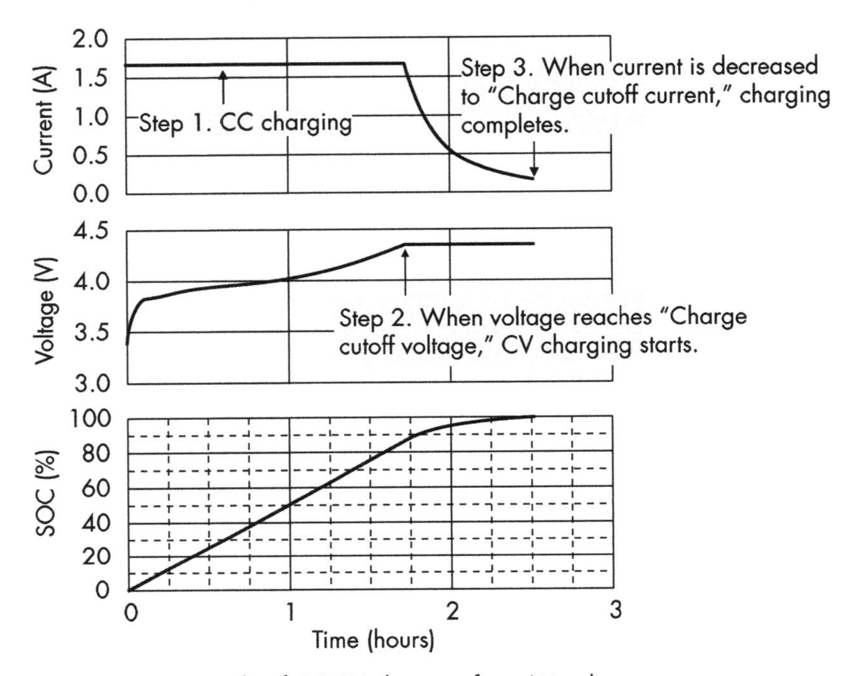


**Figure 4.1** An example of CC-CV charging for a Li-ion battery.

During CV charging, charging current is naturally reduced by the cell itself, where the details are explained in Section 4.2.3. When current is decreased to the charge cutoff current, battery charging completes. The specific charging conditions depend on the battery spec. The example in Figure 4.1 is 0.5C at CC, charge cutoff voltage and the voltage at CV are 4.35V, and charge cutoff current is 0.05C. In this case, CC charging quickly charges the battery from 0% to ~90% SOC in 1.75 hours. Then CV charging takes another 0.75 hour to charge the remaining ~10%. Why are both CC and CV needed for Li-ion battery charging? Before thinking about the reason for CC-CV charging, let's think about the impact of battery impedance during charge.

### 4.2.2 IR Jump

Chapter 3 explained that a battery has impedance. During discharge, battery voltage drops due to IR drop caused by the impedance. Then, what would happen during charge? In fact, batteries show higher voltage than OCV during charge. Figure 4.2 shows the comparison between the observed battery voltage during charge and OCV.

In this figure, the OCV curve was obtained separately by charging or discharging the battery to the specific SOCs and leaving it under no current until the battery voltage becomes stable, which is OCV.

As the figure shows, the observed voltage during charge is higher than OCV. This is because, compared to discharge, the reverse situation happens during charge. Figure 4.3 shows a schematic illustration of a battery and its internal impedance.

For convenience's sake, the impedance portion of the battery is expressed by one resistor. In reality, the impedance portion is expressed with multiple components such as resistors and parallel RC circuits, as explained in Chapter 3. During discharge in Figure 4.3(a), the impedance portion takes voltage and causes the observed voltage to look lower. In this example, 1A is supplied from the battery and the battery OCV is 3.8V. The impedance takes 1A × 0.1 ohm = 0.1V. Therefore, the observed voltage is 3.8V – 0.1V = 3.7V. During charge in Figure 4.3(b), current at 1A flows in an opposite way, thus the 0.1V change in the impedance is in the opposite direction as shown in the figure. As a result, the observed voltage is 3.8V + 0.1V = 3.9V. In this book, such voltage increase during charge is called the IR jump.

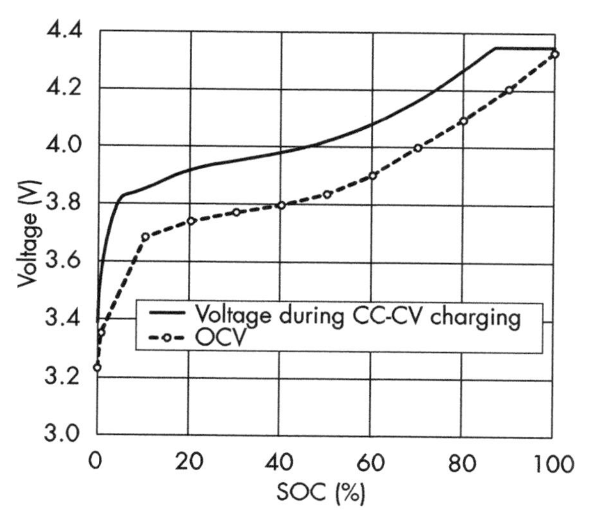


**Figure 4.2** Battery voltage during charge compared to OCV.

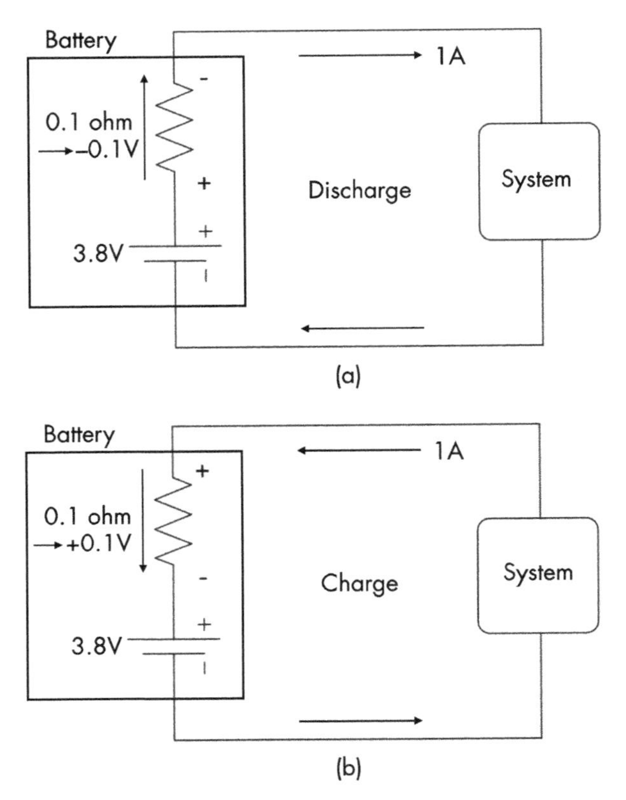


**Figure 4.3** Examples of the battery voltages during (a) discharge, and (b) charge.

***Exercise***

When a battery has 3.8V as OCV with 0.1 ohm impedance, and is charged at 2A, what is the observed voltage from the system?

***Answer***

$3.8\text{V} + 2\text{A} \times 0.1 \text{ ohm} = 4.0\text{V}$

### 4.2.3 Reason Behind CC-CV Charging

Now it is time to think again why CC-CV charging is used for a Li-ion battery. Figure 4.4 shows how OCV, battery voltage, and current change during CC-CV charging.

At the point a, the observed voltage is $V_a$. Due to the IR jump, this observed battery voltage looks higher than the actual battery voltage. Point a′ is the actual battery voltage, which is the OCV. As this

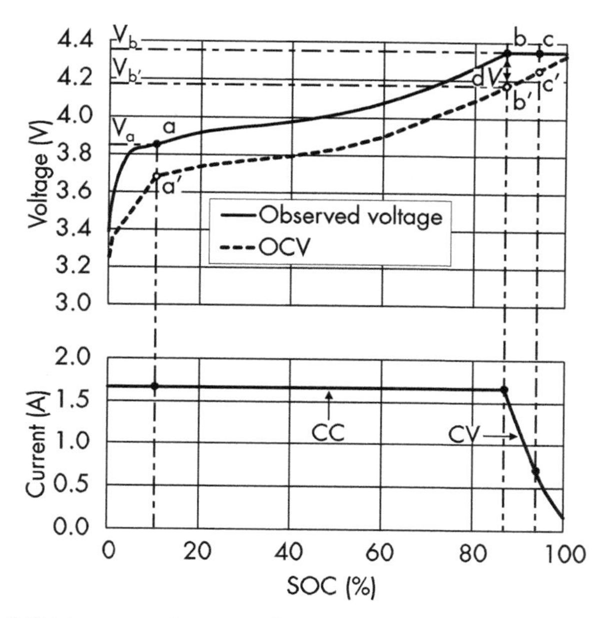


**Figure 4.4** OCV, battery voltage, and current during CC-CV charging.

battery spec allows charging to the voltage at $V_b$, which is higher than $V_a$, charging can be continued after point a. In this phase, CC charging is performed because it does not exceed the charging voltage limit in the spec. When the observed voltage reaches $V_b$ at point b, what happens if charging current stops? The battery voltage comes back to OCV ($V_{b'}$), which is the voltage at point b′. The battery still can be charged because the battery charging voltage limit is $V_b$. However, if CC resumes, the observed battery voltage immediately increases to $V_b$ due to IR jump. The voltage of the IR jump, d$V$, equals $IR$ where $I$ is charging current, and $R$ is the battery impedance. If CC is continued, the observed battery voltage exceeds $V_b$, which causes a safety risk. To charge more without exceeding $V_b$, the charging current $I$ needs to be reduced continuously. Such charging can be performed by CV charging. CV charging applies constant voltage, $V_b$ in this case. Under CV charging, the battery voltage does not exceed $V_b$, and charging current $I$ is automatically reduced by the following equation:

$$I = dV/R = (V_b - V_{OCV})/R \tag{4.1}$$

where $V_{OCV}$ is the OCV of the battery. Note that $I$ is a function of SOC because $V_{OCV}$ increases as SOC increases.

Proceeding further, when CV charging continues to point c, charging current is reduced to $(V_c - V_{c'})/R$ where $V_c$ equals $V_b$, and $V_{c'}$ is the OCV at point c′. This is lower than $(V_b - V_{b'})/R$ at point b because $V_{c'}$ is higher than $V_{b'}$.

Therefore, current is continuously and automatically reduced as CV charging goes on. Such CV charging continues until the charging current is reduced to the cutoff limit that is the charge cutoff current. If the cutoff limit is lowered, the battery can be charged slightly more. However, it takes a progressively longer time. Users may feel uncomfortable if a battery indicator of a system, such as a smartphone, stays at 99% for a long time. Therefore, the cutoff limit is set moderately low, typically somewhere between 0.1C and 0.02C, to optimize both battery capacity and charging time.

***Exercise***

If battery impedance is high, what is the impact on battery charging time?

***Answer***

A battery with higher impedance takes longer charging time. This is because the CC phase ends earlier due to a higher IR jump, and the remaining capacity must be achieved through CV charging, which is slower than CC charging.

### 4.2.4 Charging Time Simulation

With battery OCV and impedance information, it is possible to simulate battery charging time. This section explains the simulation method with exercises.

***Exercise***

You have a 3.33-Ah Li-ion battery that is empty. The battery charging spec is 0.5C CC to 4.35V, followed by 4.35V CV with 0.1C cutoff. For convenience's sake, battery impedance is constant at 0.1 ohm. Table 4.1 is an example of battery capacity and OCV relationship.

1. How long does CC charging take?

**Table 4.1**
An Example of Battery Capacity and OCV Relationship

| Charged Capacity (Ah) | OCV (V) |
|---|---|
| 3.33 | 4.33 |
| 3.30 | 4.317 |
| 3.00 | 4.2 |
| 2.93 | 4.18 |
| 1.67 | 3.835 |

2. When CV charging completes at 0.1C cutoff, what is the OCV and how much capacity is charged in total?
3. How long does it roughly take to complete CV charging?
4. What is the total charging time?

***Answer***

1. CC charging is performed at 0.5C, which is 3.33 Ah × 0.5 = 1.665A. CC charging completes when 4.35V is reached. The observed 4.35V includes IR jump, which is 1.665A × 0.1 ohm = 0.1665V. Therefore, OCV is 4.35 – 0.1665 ≈ 4.18V. This corresponds to 2.93 Ah in the table. To charge 2.93 Ah, it takes 2.93 Ah/1.665A ≈ 1.76 hours.
2. IR jump at 0.1C cutoff is 0.1C × 3.33 Ah × 0.1 ohm = 0.0333V. This means that OCV at the end of CV charging is 4.35V – 0.0333V ≈ 4.317V. This corresponds to 3.30 Ah in the table. Therefore, 3.30 Ah is charged in total.
3. When CV charging starts at 2.93-Ah charge, the charging current is 1.665A. When CV charging reaches 3.00-Ah charge, the OCV is 4.2V in the table. The charging current at 3.00-Ah charge is calculated as (4.35V – 4.2V)/0.1 ohm = 1.5A. Therefore, the average charging current from 2.93 Ah to 3.00 Ah is (1.665A + 1.5A)/2. This charges 3.00 Ah – 2.93 Ah = 0.07 Ah, which takes 0.07 Ah/[(1.665A + 1.5A)/2] ≈ 0.044 hour. The same calculation method can be applied to the CV charging from 3.00 Ah to the end of the discharging where the capacity is 3.30 Ah and the cutoff current is 0.1C (i.e., 0.333A). In this region, the capacity difference is 3.30 Ah – 3.00 Ah = 0.30

Ah. The average current is (1.5A + 0.333A)/2. To charge the 0.30 Ah, it takes 0.327 hour. In total, CV charging takes 0.044 + 0.327 ≈ 0.37 hour.

4. CC charging takes 1.76 hours and CV charging takes 0.37 hour. In total, it takes 2.13 hours.

* * *

On a side note, the measured charging time for this battery was 2.24 hours. The difference mainly comes from two reasons. One is CV charging time estimation, and the other is impedance variation. In this exercise, CV charging is segmented into two regions and each charging time is calculated by the average current. However, the current decrease in CV charging is not linear as Figure 4.1 shows. Narrower segmentation with more capacity-OCV data will make the estimation more accurate. Also in this exercise, the constant impedance was used to simplify the calculation. The real impedance depends on SOC, temperature, degradation status, and so forth. When battery temperature decreases or a battery is degraded, the impedance generally increases. This means that the transition from CC to CV starts to happen earlier in SOC. Gradually less time is spent in the CC phase and more in the CV phase. Using the appropriate impedance is another way to estimate the charging time more accurately. Also note that, when a battery is degraded and capacity is decreased, the capacity-OCV table needs to be updated. This is related to fuel gauging, which is explained in Chapter 7.

## 4.3 FAST BATTERY CHARGING

Fast battery charging is widely adapted in systems, such as smartphones, laptop PCs, and electric vehicles. It can be categorized into two approaches: continuous fast charging, and step charging. Figure 4.5 shows examples of the two fast-charging methods and normal CC-CV charging for a 3-Ah Li-ion battery. This section explains the details of each method.

### 4.3.1 Continuous Fast Charging

In Figure 4.5, the dashed line shows how current and SOC changes during continuous fast charging. For comparison, normal CC-CV

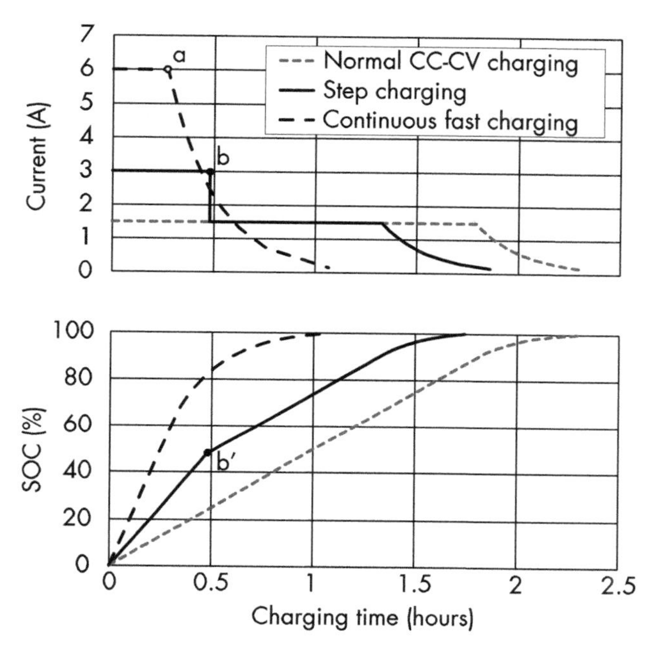


**Figure 4.5** Examples of continuous fast-charging, step charging, and normal CC-CV charging for a 3-Ah Li-ion battery.

charging is also shown with the dotted line. Continuous fast charging is also CC-CV charging, but it starts with a higher constant current than normal CC-CV charging. In Figure 4.5, CC of the continuous fast charging is 2C of a 3-Ah cell, which is 6A, while normal CC-CV starts at 0.5C, which is 1.5A. When the higher constant current reaches the charge cutoff voltage at point a, CV charging follows. As the charging current is higher than normal CC-CV charging, SOC increases faster. In this case, the continuous fast charging completes in ~1.1 hours, whereas normal CC-CV charting takes ~2.3 hours.

While continuous fast charging substantially reduces charging time, applying such high current to a battery not designed for fast charging, such as a high-energy-density battery, causes heat due to the battery impedance. That decreases cycle life and may create safety risks. To enable continuous fast charging, the battery needs to have a lower impedance. Sometimes a safer material is also used. The trade-off is lower energy density. This is because low impedance requires electrode design changes, such as increasing the thickness of cathode and anode metal sheets (i.e., current collectors, which are explained in Chapter 6). That reduces the space for the cathode and anode materials (i.e., active materials, which are explained in Chapter 6), resulting in lower energy density. A safer material, such as a $Li(Ni_{1/3}Mn_{1/3}Co_{1/3})O_2$ cathode instead of a $LiCoO_2$ cathode, can utilize less lithium ions by the nature of the material. This also results in lower energy density. This means that when the battery is capable of continuous fast charging, energy of the battery is lower than that of the same-sized battery designed for normal CC-CV charging. Selecting a proper battery design considering the requirements of battery life and charging time for the system is key to a continuous fast-charging method.

### 4.3.2 Step Charging

Step charging is a method consisting of several levels of charging current. An example is shown as solid lines in Figure 4.5. In this case, step charging starts at 3A CC, which is higher than the CC of normal CC-CV charging. Then, at point b, charging current is decreased and normal CC-CV charging follows. As the initial charging current is higher than normal CC-CV charging, SOC increases faster than normal CC-CV charging until point b′, followed by normal charging speed. This method includes a current step. Therefore, this is called step charging.

In addition to two-step CC-CV charging, explained in Figure 4.5, there are several other kinds of step charging. Examples are shown in Figure 4.6.

In this figure, three types of step charging are applied to the same battery with high-energy density. The solid line shows current, and the dotted line shows voltage.

Figure 4.6(a) is a two-step CC-CV as just explained. In this example, the first CC at 1.0C continues to 4.1V, followed by the second CC at 0.5C to 4.35V and CV at 4.35V to 0.05C cutoff. Compared to normal CC-CV charging (CC at 0.5C to 4.35V and CV at 4.35V to 0.05C cutoff), which takes 2.43 hours, this two-step CC-CV reduces charging time by 0.27 hour. During transition from 1.0C to 0.5C, battery voltage drops. This is because IR jump from OCV decreased as current decreased. If the battery allows higher than 0.5C until 4.1V, adding CV between the first and second CC reduces charging time.

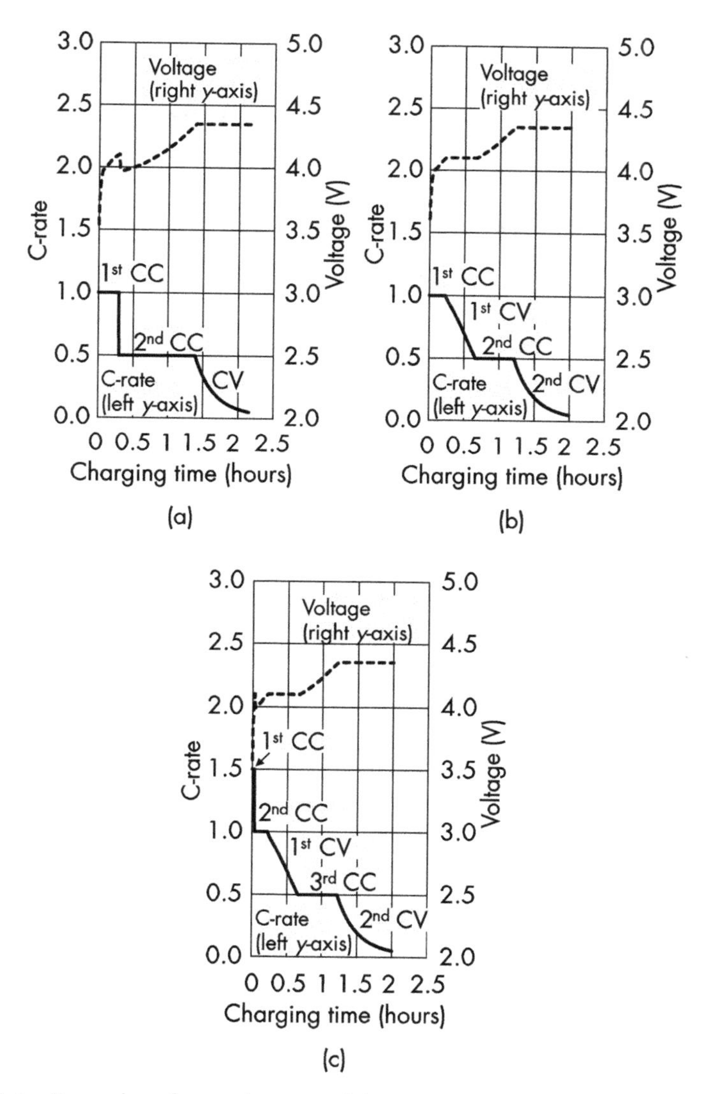


**Figure 4.6** Examples of step charging: (a) two-step CC-CV, (b) CC-CV-CC-CV, and (c) CC-CC-CV-CC-CV.

In Figure 4.6(b), the first CV at 4.1V is added between the first and second CC of two-step CC-CV in Figure 4.6(a). The first CV gradually reduces current from 1.0C to 0.5C without exceeding 4.1V. This reduced charging time by 0.16 hour, compared to Figure 4.6(a).

In Figure 4.6(c), charging starts with the even higher first CC at 1.5C to 4.1V, followed by the same charging procedure with Figure 4.6(b). As the figure shows, the first CC immediately reaches 4.1V due to large IR jump and switches to the second CC. Due to almost no contribution of the first CC, total charging time in Figure 4.6(c) is similar to that in Figure 4.6(b). This reiterates the impedance impact on charging time explained early.

If the battery is properly designed, step charging is applicable to the battery designed for continuous fast charging as well as the battery designed for normal CC-CV charging, such as a high-energy density battery. With a step charging method, the battery designed for continuous fast charging can start with even higher current than continuous fast charging. The high-energy density battery designed for normal CC-CV charging can shorten charging time with a step charging method without a trade-off against energy density. This is partly because charging at low SOC gives little negative impact on cycle life [1].

Figure 4.5 shows point b as an example of a transition point in step charging. CC current and the maximum voltage to transition to the next step depend on many factors, for example, battery materials, design, impedance, tolerable heat, desired cycle life, degradation mechanism, safety, and so forth. Point b is typically defined as the moment that the battery reaches a certain intermediate voltage. Finding a proper transition point is key to successfully developing a step charging method. Note that step charging is not always employed because the development and qualification of the method take engineering resources and costs. Also, it is a more complex charging method, typically supported only by special charger ICs or special firmware.

### 4.3.3 Fast-Charging Time Simulation

This section applies step charging knowledge to simulate the charging time.

***Exercise***

You have a 3.33-Ah Li-ion battery that is empty. Step charging is applied to the battery, where the first CC is 1C to 4.168V, the second CC is 0.5C CC to 4.35V, and CV is 4.35V with 0.1C cutoff. For convenience's sake, battery impedance is constant at 0.1 ohm. The battery capacity and OCV relationship is the same in

Table 4.1. The only difference from the exercise in Section 4.2.4 is the first CC charging.

1. How much capacity does the first CC charge and how long does it take?
2. How much time does the step charging save compared to the normal CC-CV charging in Section 4.2.4?

***Answer***

1. The first CC at 1C, which is 3.33A, is performed until 4.168V. This voltage includes IR jump. At the voltage, OCV is 4.168V – 3.33A × 0.1 ohm = 3.835V. In the table, this corresponds to 1.67 Ah. Therefore, the first CC charges 1.67Ah. It takes 1.67 Ah/3.33A ≈ 0.50h.
2. After the first CC of the step charging, 0.5C-CC and CV follows, which is the same with the exercise in Section 4.2.4. Therefore, the charging time difference comes from the charging time for the first CC. The first CC charged 1.67 Ah in ~0.50h. If this is charged at 0.5C, it takes ~1.00h. Therefore, the saved time is ~0.50h. On a side note, total charging time is estimated as 2.13h – 0.50h = 1.63 hours.

* * *

### 4.3.4 Four Key Elements for Battery Charging

To perform battery charging, four key elements in Figure 4.7 must be considered, which are battery, charging algorithm, charging circuit, and power source. These elements are important especially for fast charging.

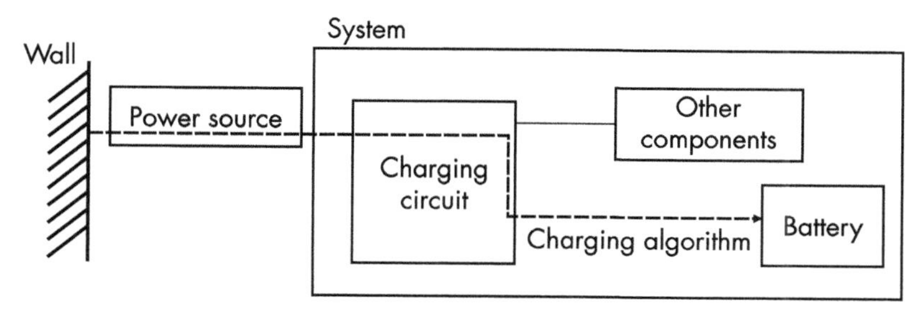


**Figure 4.7** Four key elements for battery charging.

- The battery needs to be capable of receiving the charging current, such as the fast-charging current.
- The charging algorithm must be appropriate to prevent a risk to the battery. For example, CC-CV charging, or appropriate step charging is used instead of CV charging from the beginning. There is also a charging algorithm that uses machine-learning algorithms and adjusts charging speed by predicting the user's fast charging necessity from the user's past usage data. As this can avoid unnecessary fast charging and associated heat, cycle life extension is expected. This is called context-based battery charging [2]. The details are explained in Chapter 12.
- The charging circuit needs to be capable of handling the current and voltage of the charging algorithm. This includes the considerations of the heat generation throughout the charging path and the ability of the charging system to deliver the current, such as fast charging current.
- The power source, such as a USB charger, must be powerful enough to support at least the charging current. It usually needs to be more powerful to support other components in the system (e.g., a display and a processor) simultaneously while charging.

To understand the importance of these four elements, let's think about a battery of an electric vehicle (EV) that can be charged quickly within 10 minutes. When an EV battery is 100 kWh and 10-minute charging requires at least 6C, a charger must support 600 kW that is several thousand amperes. Even if the battery can accept the current, the power supply also must support it with a thick charging cable, a large power adapter, heat management system, and so forth. Considering such cost and size is required to enable fast charging.

For portable devices, such as smartphones and laptop PCs, a USB charger is widely used. USB is an abbreviation of universal serial bus, where supply voltage and current are specified. For example, when a USB charger fully supports the USB Power Delivery (PD) Revision 3.1 spec with an appropriate USB type-C cable, the charger can deliver up to 240W, where voltage is up to 48V and current is up to 5A. A different version of USB has a different spec. For example, one of the long-standing USB specs can deliver only up to 7.5W, where voltage is 5V and current is 1.5A. This reiterates the importance of selecting the right power source for fast charging. When a smartphone has a 3-Ah battery and supports 1C charging, which requires at least 3A, a 7.5W-USB spec that provides only 1.5A is not enough. 1C charging requires a more powerful USB charger.

## 4.4 SAFE BATTERY CHARGING

### 4.4.1 Safety Guideline and Design

Battery charging under inappropriate conditions may create safety risks. To avoid that, there are industry guidelines and requirements, such as IEC62133. Figure 4.8 is an example.

Battery charging voltage and current must be within the specified voltage and current limit. Charging voltage and/or current are regulated or stopped at higher or lower temperature than the standard temperature range specified by the battery manufacturer. When battery temperature is low, ionic mobility slows down in a battery cell, and lithium ions may be plated on the surface of the anode during charge instead of being safely stored in the anode material. If lithium plating happens, dendrites may grow and cause an internal short circuit. When battery temperature is high, battery degradation is accelerated, especially when battery SOC is high. High-charging current

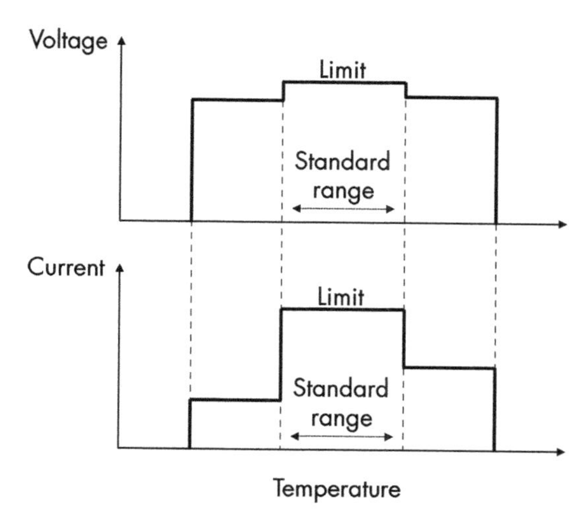


**Figure 4.8** An example of charging voltage and current limit by temperature ranges.

increases battery temperature even higher due to joule heat caused by battery impedance. To avoid such safety risks and degradation, battery charging voltage and/or current are regulated or stopped outside the standard temperature range. Battery temperature is typically monitored using a temperature sensor located near the battery.

Usually, battery charging is supposed to be performed within the limit. However, an abnormal situation, such as a malfunction in the power adapter, may happen. To avoid impact on battery safety, a battery protection device provides several protection functions, for example, over or under-voltage protection, overcurrent protection, and over- or under-temperature protection. Such protections are provided not only by a discrete protection device, such as a protection circuit module (PCM), but also by a charger IC. The details of battery protection are explained in Chapter 6.

Battery manufacturers usually specify the charging voltage and current depending on temperature ranges, and the system employs the spec for battery charging.

### 4.4.2 Precharge

The discharging voltage limit of a Li-ion battery depends on cathode and anode materials and is usually 2.75V to 3.0V in $LiCoO_2$ cathode and graphite anode. A battery may be deeply discharged below the discharging voltage limit, for example, after the battery is left unused for a while and self-discharging happens. While a user may want to charge the battery as quickly as possible, applying normal battery charging current to the battery may cause a hazardous situation because the reason for the deep discharging may be due to an internal short circuit. If that is the case, the applied charging current goes through the short-circuit path in the battery instead of moving the lithium ions and makes the situation even worse. To avoid that, there are charger ICs that provide a precharge function. Figure 4.9 is an example of precharge.

As the figure shows, precharge starts with low current when the battery is in a deeply discharged state. When battery voltage increases to the normal range within a set period, normal CC-CV charging follows. In other words, when the battery voltage does not reach the normal range within the set period, charging is terminated because the battery may have an internal short circuit.

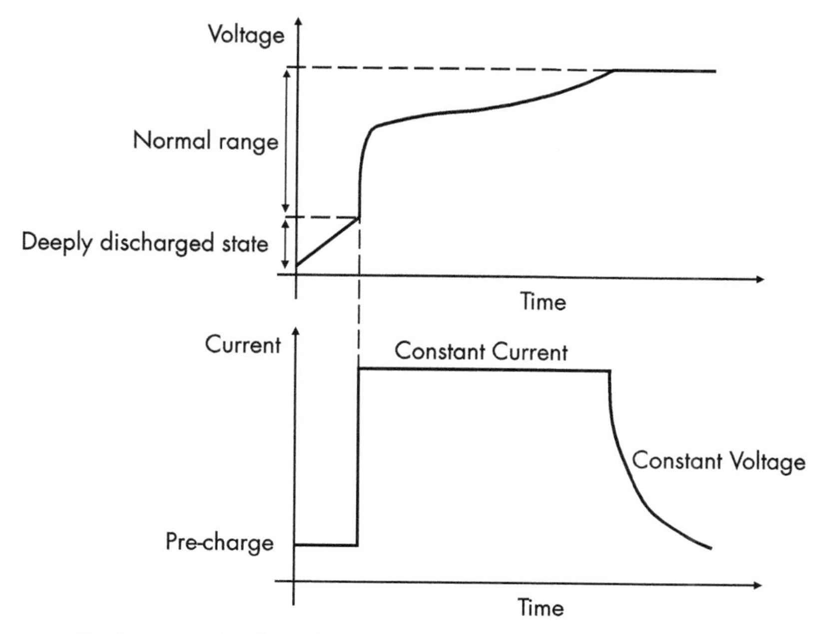


**Figure 4.9** An example of precharge.

## 4.5 WIRELESS CHARGING

### 4.5.1 Introduction

While we are on the subject of charging, it should be mentioned that many portable devices, such as smartphones, employ wireless charging as well as wired charging. Wireless charging brings its own considerations. To start wireless charging, a user just needs to place the device on a wireless charging pad. It is more convenient than wired charging where a user needs to plug and unplug a charging cable. This section explains the theory, advantages, and disadvantages of wireless charging.

### 4.5.2 Theory and Structure

Wireless charging is one method to transfer energy wirelessly. Portable devices typically use inductive charging. Figure 4.10 shows a schematic illustration of inductive charging between a system and a charging pad.

Inductive charging consists of two coils. One is a transmitting coil on a charger side, for example, a wireless charging pad. The other

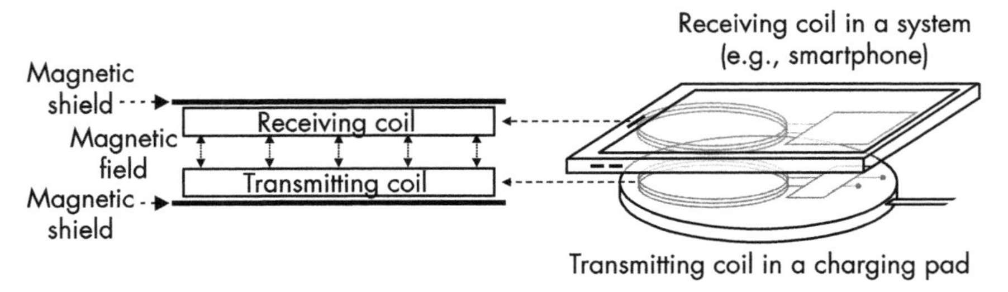


**Figure 4.10** A schematic illustration of inductive charging between a system and a charging pad.

is a receiving coil on a system side, for example, a smartphone. When an AC is supplied to the transmitting coil, an oscillating magnetic field is created because of Ampère's circuital law. When the oscillating magnetic field passes through the receiving coil, an alternating electromotive force in the form of a voltage is induced in the receiving coil by Faraday's law of induction, resulting in an AC in the receiving coil. Such AC is converted to DC through a rectifier, and supplied to a battery for charging and other components in the system.

Wireless charging consists of not only coils and control circuits, but also magnetic shields. Without the shields, the magnetic field may go through unintended areas, such as metals in a battery or a system, generate heat, and cause a safety risk.

Wireless charging is part of the power source in Figure 4.7. To support battery charging, sufficient power needs to be transferred to the charging circuit. Similar to the USB standards, there are wireless charging standards, such as Qi [3]. Qi is pronounced "chee," a Chinese word meaning energy flow. Maximum power specs depend on the revisions. It is important to make sure that the power is sufficient for battery charging. Otherwise, charging speed is constrained.

### 4.5.3 Advantages and Disadvantages

Wireless charging has advantages over wired charging, such as convenience and durability. A user only needs to place a system on a wireless charging pad. Unlike wired charging, there is no charging cable or connector that may break after repeated plug and unplug. However, there are some disadvantages, such as cost, thickness, and efficiency.

Wireless charging needs coils, shields, and controllers to transfer power wirelessly complying with a standard. These are cost disadvantages of wireless charging. Also, the coils and shields take space and increase the thickness of a system, such as a smartphone. This is not preferred from an industrial design standpoint. A possible solution is to combine them with the existing components, such as an antenna, to mitigate thickness increase. More importantly, power-transfer efficiency of wireless charging is not as high as that of wired charging due to more conversions in power delivery, resulting in more heat generation in a system. When the heat increases battery temperature above the standard temperature range, battery charging is limited. The heat also affects system performance as IC may slow down not to exceed its thermal limit. The heat will also accelerate battery degradation. In the case of a small system with a small battery, such as watches and smartphones, the impact may be small because of low power. However, in the case of a large system with a large battery, such as electric vehicles, inefficiency from high power is more challenging. Therefore, mitigation plans need to be considered, such as better thermal management and charging current reduction.

### 4.5.4 Essentials of Wireless Charging for Battery Engineers

Wireless charging is one of the power delivery methods and may not be the area readers of this book work on. However, knowing the advantages and disadvantages of the technology sparks the creativity in the collaboration with those engineers. If you are a battery engineer for the system that uses wireless charging, it is important to know how much charging power is available to the battery and how much heat is transferred to the battery.

## 4.6 SUMMARY

In this chapter, we learned the following:

- During charge, battery voltage looks higher (i.e., IR jump) due to impedance.
- Impedance affects charging time.

- Fast charging can be categorized into two methods: continuous fast charging, which is typically a trade-off with energy density, and step charging, which can increase charging speed without affecting energy density but requires more complex controls.
- Battery charging requires four key elements: battery, charging algorithm, charging circuit, and power source.
- Wireless charging requires consideration of cost, thickness, and heat.

## 4.7 PROBLEMS

### Problem 4.1

You have a 3.3-Ah Li-ion battery where the battery OCV is 3.4V, DC impedance is 120 mohm, safety limit of charging current is 3.3A, and CV charging voltage is 4.35V. What happens if CV charging instead of CC charging is applied to the battery?

### Answer 4.1

(4.35V – 3.4V)/0.12 ohm = ~7.9A is supplied. This is above the safety limit of charging current. Therefore, the battery charging should start with CC charging.

### Problem 4.2

You bought a new smartphone that includes a 19-Wh battery. The battery is nominal 3.8V. The smartphone claims a quick charging feature that charges the battery from 0% to 50% SOC in 15 minutes at constant current. One day, you need to charge the smartphone with a different charger where the output is 2.1A at 5V. What would happen?

### Answer 4.2

50% charging in 15 minutes at constant current means that the battery is capable of 2C charging. As the battery is 19 Wh and nominal voltage is 3.8V, capacity is 19 Wh/3.8V = 5 Ah. 2C of 5Ah is 10A. The charger provides 2.1A at 5V, which is not sufficient to enable 2C charging. Therefore, 2C charging speed cannot be expected with the charger.

Problem 4.3

For the smartphone and the battery in Problem 4.2, a genuine charger that supports 2C charging is used. However, slower charging is observed. Explain the possible reasons.

Answer 4.3

Example 1: The smartphone may be powered on and consume power for display, CPU, and so forth, in addition to battery charging. In this case, input power may be shared, resulting in slower battery charging.

Example 2: The temperature of the smartphone and the battery may be outside the standard temperature range. In such a situation, battery charging current may be limited.

Example 3: There may be poor contact made to the charger, which increases impedance through the charging path. It is good to check contact points to be sure they are clean and free of corrosion.

Problem 4.4

You have a Li-ion battery that supports step charging. The spec shows that the step charging can start at up to 2C CC until 4.1V, followed by 0.5C CC and CV charging. When the battery was charged from empty at 2C CC, the 2C charging stopped in four minutes at 15% SOC, followed by 0.5C CC and CV charging. The total charging time was two hours. What was the possible reason for 2C stopping in four minutes and what is the possible solution to make the total charging time shorter?

Answer 4.4

While charging, an IR jump happens, 2C charging is high current that causes a large IR jump and may soon reach the charge cutoff voltage of the first CC, which is 4.1V. If that is the case, one possible option to shorten the charging time is to add 4.1V CV charging in between the first CC at 2C and the second CC at 0.5C. When the first CC reaches 4.1V and switches to CV charging at 4.1V, charging current is gradually decreased from 2C without exceeding

4.1V. When the current is decreased to 0.5C, the second CC can follow. As the current of CV charging at 4.1V is higher than 0.5C, the charging time can be shorter. This is one possible solution. It is always important to confirm that the charging meets the spec and can be performed safely.

## References

[1] Takei, K., et al., "Cycle Life Estimation of Lithium Secondary Battery by Extrapolation Method and Accelerated Aging Test," *Journal of Power Sources*, Vol. 97–98, 2001, pp. 697–701.

[2] Matsumura, N., "Context-Based Battery Charging Algorithm, an Application of Machine Learning/Deep Learning to Battery Charging for Longevity Extension," *The 39th International Battery Seminar & Exhibit*, Florida, 2022.

[3] https://www.wirelesspowerconsortium.com/.

# 5

# PRESENT AND FUTURE BATTERIES

## 5.1 INTRODUCTION

Li-ion rechargeable batteries have changed our lifestyle. Thanks to the high gravimetric and volumetric energy density of Li-ion batteries, smartphones and laptop PCs provide long battery life in stylish and lightweight designs, and electric vehicles provide long driving range. There are many other rechargeable batteries, such as lead-acid batteries and Ni-MH batteries. Even within Li-ion batteries, there are many kinds of cathodes and anodes. This chapter explains the chemical reactions and advantages of present and future rechargeable batteries.

### 5.1.1 Introduction of Rechargeable Batteries

Figure 5.1 shows the comparison of energy density among popular rechargeable batteries in the market.

In this chart, the *x*-axis is volumetric energy density, which is Wh/l, and the *y*-axis is gravimetric energy density (or specific energy), which is Wh/kg. The plots are examples from commercially available Li-ion, Ni-MH, and lead-acid batteries. Ni-MH is read as nickel metal hydride. Li-ion batteries provide far greater energy density than others and are widely used in portable systems, electric vehicles, and elsewhere.

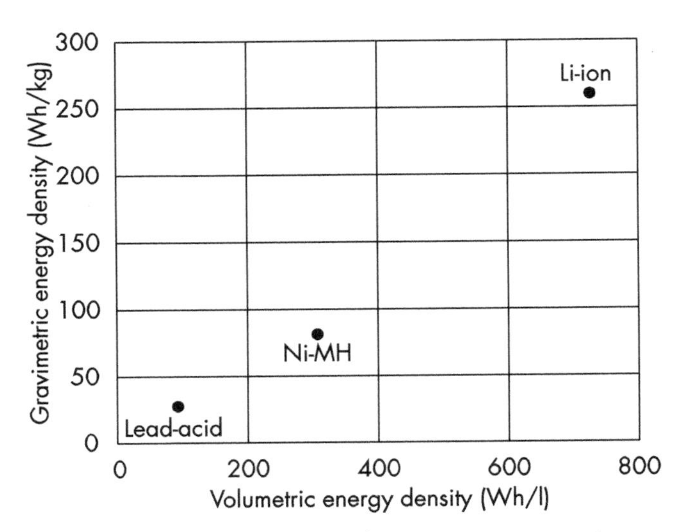


**Figure 5.1** An example of volumetric and gravimetric energy density comparison among rechargeable batteries.

Note that rechargeable batteries are sometimes referred to as secondary batteries, in contrast to nonrechargeable batteries known as primary batteries.

### 5.1.2 Rechargeable Battery Usage

Li-ion batteries provide the highest energy density but do not dominate the systems that need rechargeable batteries. For example, lead-acid batteries are used as backup batteries because of their reliability and low cost, and also are used in cars for SLI, which means starting, lighting, and ignition. Ni-MH batteries are also used as rechargeable alternatives to nonrechargeable alkaline batteries such as AA and AAA batteries. Each battery technology has its own advantage.

## 5.2 LEAD-ACID BATTERY

### 5.2.1 Reactions

A lead-acid battery consists of a lead dioxide cathode, $PbO_2$, and a lead anode, Pb, in an electrolyte that consists of water and sulfuric acid. Chemical reactions are expressed as follows:

Cathode:

$$PbO_2 + 4H^+ + 2e^- = Pb^{2+} + 2H_2O \tag{5.1}$$

$$Pb^{2+} + SO_4^{2-} = PbSO_4 \tag{5.2}$$

Anode:

$$Pb = Pb^{2+} + 2e^- \tag{5.3}$$

$$Pb^{2+} + SO_4^{2-} = PbSO_4 \tag{5.4}$$

Overall:

$$Pb + PbO_2 + 2H_2SO_4 = 2PbSO_4 + 2H_2O \tag{5.5}$$

Figure 5.2 shows a schematic illustration of a lead-acid battery during discharge.

As shown in the figure and reactions (5.1–5.4), two reactions happen in both cathode and anode, forming $PbSO_4$ during discharge. The OCV of the overall reaction (5.5) is 2.1V [1]. The voltage of a car battery for SLI is typically above 12V. This is enabled by having six lead-acid battery cells connected in series.

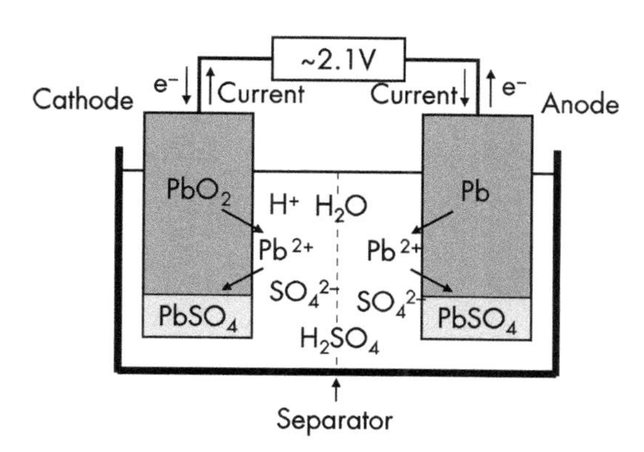


**Figure 5.2** A schematic illustration of a lead-acid battery during discharge.

### 5.2.2 Advantages and Disadvantages

The advantages of a lead-acid battery over a Li-ion battery are cost, recyclability, and safety. It is low cost because there is no expensive material in a lead-acid battery, unlike lithium and cobalt in a Li-ion battery. The recycling process for a lead-acid battery is well established with up to 98% of the batteries able to be recycled, while the recycling rate of a Li-ion battery is less than 5% [2, 3]. The electrolyte in a lead-acid battery is water-based, which is safer than the flammable electrolyte in a Li-ion battery. This reduces fire risk, which is important especially when the battery is used as a backup battery in a building and provides high current.

In contrast, the gravimetric and volumetric energy densities of a lead-acid battery are low as shown in Figure 5.1. Compared to a Li-ion battery, a lead-acid battery provides around one-tenth of energy density. This means that a lead-acid battery needs ten times more space and weight than a Li-ion battery to provide the same energy.

On a side note, a car battery, which is a lead-acid battery, is sometimes discharged too much, and a jump start is needed with the help of a healthy battery in another car. The instruction of the jumper cables typically reads as follows: Attach one end of the positive (red) cable to the positive terminal of the dead battery and the other end to the positive terminal of the healthy battery. Next, attach one end of the negative (black) cable to the negative terminal of the healthy battery. Then, attach the other end of the negative cable to the unpainted metal surface of the dead car. Why is the negative cable not attached to the negative terminal of the dead car? This is because hydrogen gas may be present at the negative terminal as a result of water decomposition in the past operations. Direct attachment to the negative terminal may cause sparks and an explosion. Instead, it is safer to attach the cable to the unpainted metal surface of the car, which is the common ground with the negative terminal.

## 5.3 NI-MH BATTERY

### 5.3.1 Reactions

Nickel-metal hydride battery, Ni-MH battery, consists of a nickel oxide hydroxide cathode, NiOOH, and metal hydride anode, MH, in an electrolyte that consists of water and alkaline hydroxide. M is hydrogen storage metal/alloy that stores hydrogen after charging.

Chemical reactions are expressed as follows:

$$\text{Cathode: } NiOOH + H_2O + e^- = Ni(OH)_2 + OH^- \tag{5.6}$$

$$\text{Anode: } MH + OH^- = M + H_2O + e^- \tag{5.7}$$

$$\text{Overall: } MH + NiOOH = M + Ni(OH)_2 \tag{5.8}$$

Figure 5.3 shows a schematic illustration of a Ni-MH battery during discharge. Nickel oxide hydroxide in the cathode reacts with a hydrogen ion of water in the electrolyte, receives an electron, and forms nickel hydroxide, $Ni(OH)_2$. In the anode, metal hydride releases a hydrogen ion and electron, and becomes metal. The OCV is around 1.35V [1]. While the voltage is slightly lower than 1.5V of a nonrechargeable alkaline battery, there are many systems that can accept the voltage of a Ni-MH battery. Therefore, a Ni-MH battery is used as a rechargeable alternative to an alkaline battery.

### 5.3.2 Advantages and Disadvantages

The systems that use a nonrechargeable alkaline battery can use a Ni-MH battery without hardware change because of the voltage simi-

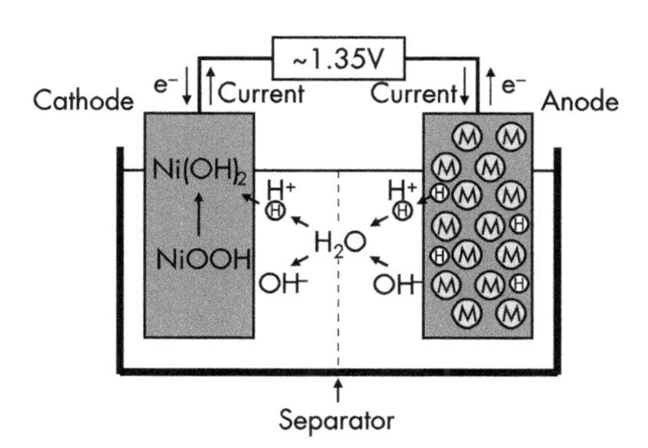


**Figure 5.3** A schematic illustration of a Ni-MH battery during discharge.

larity. As a Ni-MH battery is rechargeable and more affordable than replacing an alkaline battery, it reduces cost of ownership, which is the required total cost while the system is owned. It also enhances sustainability because of less consumption of natural resources. Its relatively lower internal resistance over nonrechargeable alkaline batteries allows it to support high-power devices. Still, its energy density is not as good as the energy density of a Li-ion battery as shown in Figure 5.1.

## 5.4 LI-ION BATTERY

### 5.4.1 Cathode and Anode Options

Li-ion batteries are widely used for systems that require high-energy density, such as smartphones, laptop PCs, and electric vehicles. When $LiCoO_2$ is the cathode and graphite is the anode, the reactions are expressed as follows:

$$\text{Cathode: } Li_{1-x}CoO_2 + xLi^+ + xe^- = LiCoO_2 \tag{5.9}$$

$$\text{Anode: } LiC_6 = C_6 + Li^+ + e^- \tag{5.10}$$

$$\text{Overall: } Li_{1-x}CoO_2 + xLiC_6 = LiCoO_2 + xC_6 \tag{5.11}$$

where $x$ is the utilization rate of lithium ions in $LiCoO_2$ $(0 \leq x \leq 1)$ such as 0.65. The utilization rate has been increasing, depending on the technology generations, but cannot be 1 as explained in Chapter 1.

There are several other cathodes and anodes as shown in Figure 5.4 [4–18].

### 5.4.2 Details of Cathodes: LCO, NMC, NCA, and LFP

In Li-ion batteries, examples of popular cathode materials are $LiCoO_2$, $Li(Ni_xMn_yCo_z)O_2$, $Li(Ni_xCo_yAl_z)O_2$, $LiFePO_4$, and $LiMn_2O_4$.

Lithium cobalt oxide, $LiCoO_2$, has a layered structure as shown in Figure 5.5.

The layered structure enables high two-dimensional conductivity of lithium ions with structural stability, resulting in good cycle life

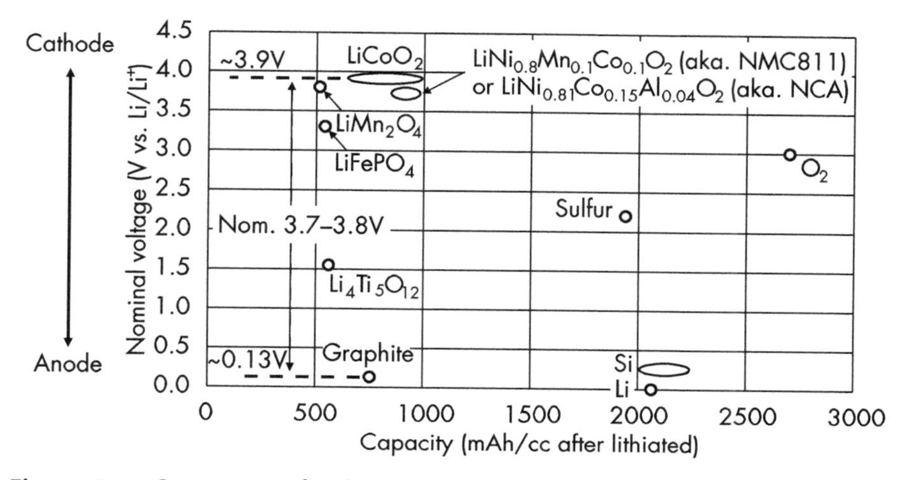


**Figure 5.4** Comparison of cathode and anode materials for Li-ion batteries. [4–18]

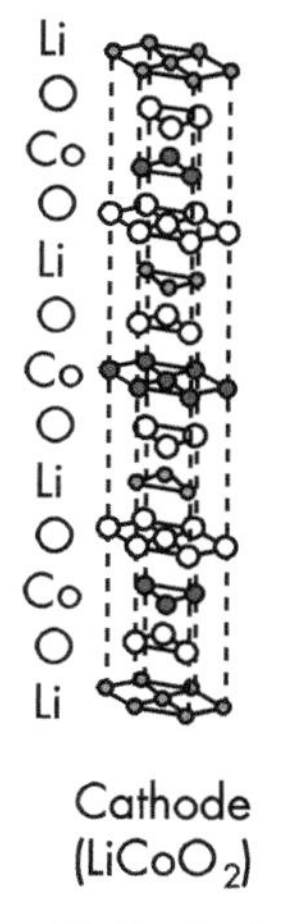


**Figure 5.5** A schematic illustration of $LiCoO_2$ layered structure.

[19]. As $LiCoO_2$ provides both high voltage and high capacity among the cathode materials, it is used in consumer electronics batteries, such as those for the smartphone and laptop PC, where high volumetric energy density is needed. $LiCoO_2$ is abbreviated as LCO.

Lithium nickel manganese cobalt oxide, $Li(Ni_xMn_yCo_z)O_2$, where $x + y + z = 1$, also has a layered structure [20]. The ratio of lithium, metals (Ni + Mn + Co), and oxygen is 1-1-2, which is the same as $LiCoO_2$. For $x$, $y$, and $z$ in $Li(Ni_xMn_yCo_z)O_2$, there are several popular ratios such as $(x, y, z) = (1/3, 1/3, 1/3)$, (0.6, 0.2, 0.2), and (0.8, 0.1, 0.1). These are abbreviated as NMC111, NMC622, and NMC811, respectively. For example, $Li(Ni_{1/3}Mn_{1/3}Co_{1/3})O_2$ is NMC111. Sometimes NMC is abbreviated as NCM instead. When the nickel ratio increases, volumetric capacity (mAh/cc) increases. Among these three cathodes, NMC811 provides the highest mAh/cc, but the trade off is lower thermal stability and cycle life. While both LCO and NMC811 are actively researched and capacity is being increased, the nominal voltage of LCO is slightly higher than that of NMC811, as shown in Figure 5.4. Therefore, LCO provides higher energy density. However, as the price of cobalt is high and unstable, NMC, which uses less cobalt, is considered an affordable cathode in Li-ion batteries.

Lithium nickel cobalt aluminum oxide, $Li(Ni_xCo_yAl_z)O_2$, where $x + y + z = 1$ also a layered structure [20]. Typically, $x$ is >0.8 and $y$ is >0.1. For example, $(x, y, z)$ = (0.81, 0.15, 0.04) and (0.84, 0.12, 0.04) [6, 21]. This chemistry is abbreviated as NCA. NCA shows similar volumetric capacity and voltage to NMC811. As NCA uses less cobalt than LCO, it is considered an affordable cathode in Li-ion batteries as well as NMC.

Both NCA and NMC are preferred in a system that needs both high-energy density and affordability, such as an electric vehicle, which is equipped with a large thus high-cost Li-ion battery.

Lithium iron phosphate, $LiFePO_4$, has an olivine structure [20]. It is abbreviated as LFP. As shown in Figure 5.4, both volumetric capacity and voltage of LFP is lower than that of LCO, NMC811, or NCA. However, LFP provides better stability at high temperatures, resulting in better safety [22]. Also, LFP does not use expensive metals, which provides more affordability than NMC or NCA. Because of these advantages, LFP lowers the battery cost of electric vehicles at the expense of reduced driving range, compared to NMC or NCA.

Lithium manganese oxide, $LiMn_2O_4$, has a spinel structure [19]. It is abbreviated as LMO. While the voltage is at a similar level to LCO, NMC811, and NCA, the volumetric capacity is lower as shown in Figure 5.4, resulting in lower volumetric energy density than LCO, NMC811, or NCA. However, as it does not contain expensive materials, such as cobalt and/or nickel, the cost is typically lower than that of LCO, NMC811, or NCA.

Table 5.1 is the summary of key cathode materials in Li-ion batteries.

### 5.4.3 Details of Anode: Silicon Versus Graphite

Graphite anode has been used in most of the Li-ion batteries in the market. After charging, lithium ions are safely stored in the layered structure of graphite as explained in Chapter 1. With the increased demand for longer battery life in systems, new anodes with higher capacity than graphite are desired. Silicon is one of the next-generation anode materials. After charging, silicon can be lithiated to $Li_{4.4}Si$ [12]. Compared to this, graphite, which is a conventional anode material, can be lithiated to $LiC_6$ after charging, as discussed in Chapter 1. This means that one silicon atom can receive 4.4 lithium atoms, while six carbon atoms are needed to receive only one lithium.

Half reaction of the silicon anode is as follows:

$$Li_{4.4}Si = 4.4Li^{+} + Si + 4.4e^{-} \tag{5.12}$$

Based on what Chapter 2 explained, let's calculate the capacity of silicon anode.

***Exercise***

When silicon is 28.1g/mol and reacts with lithium in (5.12), how much Ah/g can silicon provide? Use 96485 C/mol for the Faraday constant.

**Table 5.1**
Comparison of Cathode Materials for Li-ion Batteries

| Formula | Abbreviation | Nominal Voltage | mAh/cc | Wh/l | Cost |
|---|---|---|---|---|---|
| $LiCoO_2$ | LCO | Highest | High | Highest | Highest |
| $Li(Ni_{0.81}Co_{0.15}Al_{0.04})O_2$ | NCA | High | High | High | Medium |
| $Li(Ni_{0.8}Mn_{0.1}Co_{0.1})O_2$ | NMC811 | High | High | High | Medium |
| $LiFePO_4$ | LFP | Low | Low | Low | Low |
| $LiMn_2O_4$ | LMO | High | Low | Low | Low |

***Answer***

1 mole of silicon can take 4.4 mol lithium, which brings electrons of 4.4 mol × 96485 C/mol. 1 Ah is 3600C. Therefore, 4.4 mol × (96485 C/mol) / 3600 / 28.1 ≈ 4.2 Ah/g.

* * *

As a reference, graphite is 12.0 g/mol as carbon and provides only 1/6 mol × (96485 C/mol) / 3600 / 12.0 ≈ 0.372 Ah/g. This proves how great potential the silicon anode has gravimetrically.

From a volumetric perspective, when silicon is 2.3 g/cc (grams per cubic centimeter) and graphite is 2.2 g/cc, silicon provides 4.2 Ah/g × 2.3 g/cc ≈ 9.7 Ah/cc, whereas graphite provides 0.372 Ah/g × 2.2 g/cc ≈ 0.82 Ah/cc. It may seem that silicon anode provides 9.7/0.82 ≈ 12 times greater volumetric capacity than graphite. However, the comparison is not true. After being charged to $Li_{4.4}Si$, the silicon swells by around four times [23]. After discharge, the lithiated silicon shrinks to around one-fourth. Such substantial volume change is because silicon anode does not have a layered structure like graphite, and therefore does not have room to store lithium and must expand. Such volume change causes electrode degradations. For example, silicon particles break up and delaminate with cycling. Also, the solid-electrolyte interface (SEI) layer, which is a lithium-electrolyte compound on silicon anode particles, will crack upon expansion, creating fresh surfaces of silicon and leading to new SEI formation. That consumes lithium as an irreversible capacity loss. These degradations result in poor cycle life. In contrast, graphite swells by only 10% after charging [24]. When a battery is designed for a system, appropriate clearance for the swelling needs to be allocated in advance. Therefore, when volumetric capacities are compared, such volume expansion after charging needs to be considered. The volumetric capacities of silicon and graphite in Figure 5.4 are based on the practical values after charging (i.e., lithiated). Still, even considering the volume expansion, silicon anodes potentially provide three-times greater mAh/cc than graphite. To enable such high capacity, researchers are investigating solutions to the cycle-life challenge due to volume change during charge and discharge. While there is no industry-wide consensus solution yet that completely replaces today's graphite anode, many approaches are being researched, including nanostructures, core-shell structures, and surface coatings, among others [25].

Note that discharge cutoff voltage of some Li-ion batteries with silicon anode and $LiCoO_2$ cathode is 2.5V to 2.7V to fully access all of the capacity. This is lower than the 3.0V of typical Li-ion batteries with graphite anode and $LiCoO_2$ cathode because of the chemical potential difference between silicon and graphite. If system shutdown voltage is higher than discharge cutoff voltage, additional electronics such as a step-up voltage regulator may be required, which adds cost and volume. Serial connection of the cells is another method to boost the battery voltage, which is explained in Chapter 6.

### 5.4.4 Details of Anode: Lithium Metal

Lithium metal anode provides high capacity and 0V versus $Li/Li^+$ as shown in Figure 5.4. Because the battery voltage is the difference between the cathode and anode voltages, 0V in anode provides the highest battery voltage in the combination with any cathode materials.

Figure 5.6 shows a schematic illustration of Li-ion battery with lithium metal anode before charging (i.e., after assembled) and after charging.

In the figure, the cross-sectional illustrations of cathode with aluminum current collector and anode with copper current collector are shown. When the battery is assembled, or before charging in

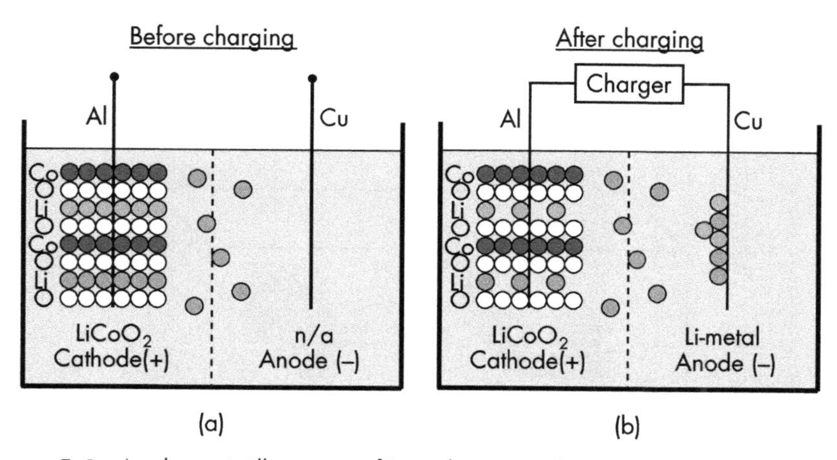


**Figure 5.6** A schematic illustration of Li-ion battery with $LiCoO_2$ cathode and lithium metal anode: (a) before charging, and (b) after charging.

Figure 5.6(a), there is no lithium in the anode. Theoretically, no or a little lithium metal is needed in the anode when the battery is assembled. This is because lithium can come from the cathode during charge and is plated on the copper anode current collector as shown in Figure 5.6(b). This leads to battery cost reduction because there is no anode material required for manufacturing. In some cases, there may be a lithium film on the anode current collector before charging for uniform lithium deposition and compensation for consumed lithium to form SEI.

When lithium is 6.94 g/mol and the Faraday constant is 96485 C/mol, it provides 96485 C/mol / 3600 / 6.94 ≈ 3.86 Ah/g as gravimetric capacity. When lithium is 0.534 g/cc, it provides 3.86 Ah/g × 0.534 g/cc ≈ 2.06 Ah/cc as volumetric capacity. This is ~2.7 times greater than today's graphite anode.

During charge, the lithium metal anode swells due to lithium plating. If a battery is designed as 2.06 Ah, the lithium metal anode swells by at least 1 cubic centimeter after full charging, assuming uniformly dense lithium is formed. The system, battery, and electrode must be properly designed to manage the swelling.

Then, is it possible to use the lithium metal anode when the clearance for the swelling is allocated? Unfortunately, using lithium metal as anode also creates a safety risk. When the Li-ion battery was first invented, metallic lithium anode was used. However, it was found that lithium plating on the anode during charge does not grow flat but generates dendrites as shown in Figure 5.7.

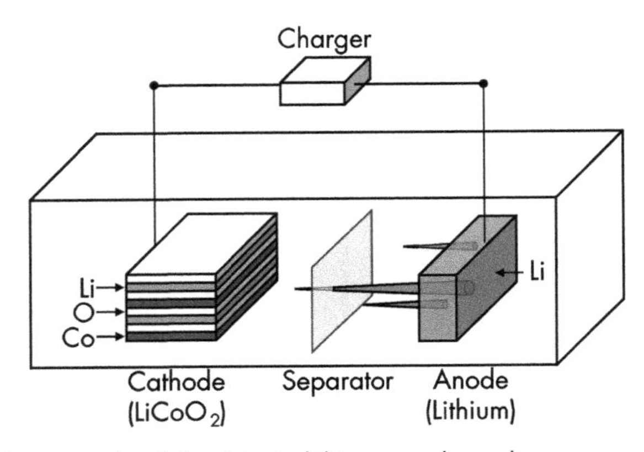


**Figure 5.7** An example of dendrite in lithium metal anode.

Such dendrites may cause internal short circuits. Because of this, graphite, which can safely store lithium in its layered structure, is still the dominant anode material today. However, lithium metal anode is attractive from the energy density standpoint. One of the potential solutions is to use a solid-state electrolyte instead of conventional liquid electrolyte. The next section explains the details of this approach.

### 5.4.5 All-Solid-State Battery

An all-solid-state battery is literally a battery that consists of solid materials without liquid. In this battery, a solid-state electrolyte (SSE) is used instead of a conventional liquid electrolyte. An SSE conducts lithium ions through its solid structure and is electronically insulative. Therefore, it replaces not only the conventional electrolyte but also the separator. Figure 5.8 is a cross-sectional illustration of an all-solid-state Li-ion battery during charge.

In this battery, during charge, lithium ions from the cathode hop through the lithium-ion transport channels in the SSE and move to the anode. During discharge, lithium ions hop back to the cathode. The battery was demonstrated early as a thin-film battery [26]. Because the SSE is mechanically harder than a conventional separator, it reduces the risk of dendrite penetration when a lithium metal anode

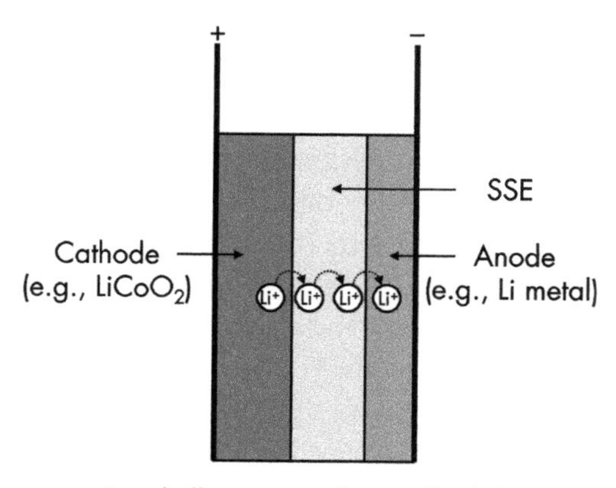


**Figure 5.8** A cross-sectional illustration of an all-solid-state Li-ion battery during charge.

is used. It also reduces the fire risk because there is no flammable electrolyte. Because of these advantages, all-solid-state batteries are expected not only to provide higher energy density with lithium metal anode, but also to provide better safety by replacing conventional flammable electrolytes.

While this technology was commercialized as a thin-film battery with small capacity, there are challenges to scale to portable systems (smartphones, laptop PCs, etc.) and electric vehicles.

*Impedance:* While ionic conductivity of the conventional liquid electrolyte in a Li-ion battery is ~$10^{-2}$ S/cm at 25°C, many of the SSE materials show lower ionic conductivity, such as ~$10^{-6}$ S/cm for lithium phosphorus oxynitride (LiPON) [27]. S in the S/cm is siemens, the unit of electrical conductance and the reciprocal of ohm (1/ohm). When the ionic conductivity is low, battery impedance and IR drop increase, resulting in less usable capacity as explained in Chapter 3. There are several SSEs that provide high-ionic conductivity. For example, $Li_{10}GeP_2S_{12}$ was reported to provide similar conductivity to the conventional electrolyte [27]. Further investigations of such materials with high conductivity and wide-operating temperature are ongoing.

*Cost:* For a thin-film form factor of all-solid-state batteries, a vacuum process such as evaporation or sputtering is used. The process enables good contact between cathode, SSE, and anode in Figure 5.8, resulting in easier hopping of lithium ions through solid interfaces. It also enables thin SSE, which reduces impedance. However, compared to today's process, which consists of electrodes-separator assembly and electrolyte injection, a vacuum process takes longer time and costs more. There are some approaches that leverage the existing manufacturing process of Li-ion batteries and replace the electrolyte with SSE. Still, it is challenging to establish good contact between SSE and the electrodes with low impedance. The details of the manufacturing process for a conventional Li-ion battery and an all-solid-state battery are explained in Chapter 6.

*Cycle life:* During charge and discharge, the materials in a cell, especially lithium metal, swell and shrink. Such volume change causes a gap in solid interfaces between cathode/anode and SSE, resulting in cycle-life degradation.

Many researchers are investigating the solutions to these challenges to scale all-solid-state batteries to portable systems and electric vehicles. Similar to silicon anode, while there is no consensus solution yet that completely replaces today's electrolyte and graphite anode, many approaches such as lithium surface coatings with SSE are being researched [25].

### 5.4.6 Details of Anode: LTO

While most of the Li-ion batteries in the market have been using graphite anode, lithium titanate ($Li_4Ti_5O_{12}$), which is abbreviated as LTO, has also been used in some specific areas. LTO has a spinel structure and is charged via the following equation:

$$Li_4Ti_5O_{12} + 3Li^+ + 3e^- = Li_7Ti_5O_{12} \tag{5.13}$$

The nominal voltage is 1.55V versus Li/Li$^+$, which is high compared to other anode materials. This means that the cell voltage in combination with a cathode material is low. Also, the volumetric capacity is lower than other anode materials. While these issues lead to low volumetric energy density, LTO has excellent cycle life because of almost no volume change during charge/discharge. It can also be charged very fast [28]. Because of these advantages, LTO has been used in electric buses, stationary storage applications, and elsewhere.

## 5.5 SUMMARY

In this chapter, we learned the following:

- Chemical reactions and advantages of key rechargeable batteries.
- Cathode and anode materials of Li-ion batteries.
- Potential of all-solid-state Li-ion batteries.

Even if different cathode and/or anode materials emerge in the future, such as a lithium-sulfur battery, an aluminum-ion battery, or a sodium-ion battery, readers can compare the new batteries with today's Li-ion battery by investigating the material properties such as mAh/cc and nominal voltage.

## 5.6 PROBLEMS

Problem 5.1

You are a battery researcher and have developed a new cathode with 1 Ah/cc and a new anode with 2 Ah/cc. The nominal battery voltage of the new cathode and anode is 4.0V. A big smartphone manufacturer contacted you and asked "Based on today's chemistry (cathode 0.8 Ah/cc, anode 0.8 Ah/cc, nominal 3.8V), we know that 11.4 Wh is possible for the size of the next-generation smartphone battery. How many Wh is roughly possible with your chemistry?" You need to explain your estimation immediately. If your new cathode and anode can replace today's chemistries without changing the cell structure, and the Ah/cc of your new cathode/anode stays the same, answer the following problems.

1. How much cathode and anode space is required based on today's chemistry?

2. If all the cathode and anode space can be used with your new cathode and anode, how many Wh is possible?

Answer 5.1

1. The capacity of the battery with today's chemistry is 11.4 Wh/3.8V = 3.0 Ah. Today's chemistry is 0.8 Ah/cc for both cathode and anode. Therefore, cathode requires 3.0 Ah/(0.8 Ah/cc) = 3.75 cc at least. Anode also requires 3.0 Ah/(0.8 Ah/cc) = 3.75 cc at least.

2. For cathode and anode, there is 3.75 cc × 2 = 7.5 cc in total. This space needs to be distributed to your new cathode and anode. Theoretically the cathode and anode must be capable of the same Ah. This is because, for example, when a cathode provides 1 Ah, anode must receive 1 Ah. Therefore, a cathode requires 7.5 cc × [(2 Ah/cc) / (1 Ah/cc + 2 Ah/cc)] = 5 cc. When your new cathode has 5 cc, 1 Ah/cc × 5 cc = 5 Ah is possible. For anode, 7.5 cc × [(1 Ah/cc) / (1 Ah/cc + 2 Ah/cc)] = 2.5 cc. When your new anode has 2.5 cc, 2 Ah/cc × 2.5 cc = 5 Ah is possible. When both cathode and anode are capable of 5 Ah, the cell is also capable of 5 Ah. As the nominal voltage is 4.0V, 5 Ah × 4.0V = 20 Wh is theoretically possible with your new cathode and anode.

On a side note, in a real battery design, cathode and anode may not be capable of the same Ah for safety reasons. For example, there may be more graphite anode than the theoretical requirement to make sure that all lithium ions can be safely stored in the graphite layers during charge.

Problem 5.2

Si anode in a Li-ion battery can be charged to $Li_{22}Si_5$ (= $Li_{4.4}Si$). However, it swells a lot. To reduce swelling, Si is charged to $Li_{13}Si_4$.

1. When Si is charged to $Li_{13}Si_4$, how much Ah/g can be provided when the density of Si is 28.1 g/mol and the Faraday constant is 96485 C/mol?

2. When Si is 2.33 g/cc and swells by 3.5 times after charging, how much is the actual mAh/cc?

Answer 5.2

1. One mole of Si can receive 13/4 moles of lithium, which is 13/4 moles of electrons. Therefore, 13/4 × (96485 C/mol) / 3600 C / (28.1 g/cc) ≈ 3.10 Ah/g.

2. (3.10 Ah/g) × (2.33 g/cc) / 3.5 = 2.06 Ah/cc

## References

[1] Thackeray, M., "Batteries, Transportation Applications," in *Encyclopedia of Energy*, pp. 127–139, C. Cleveland (ed.), New York: Elsevier, 2004.

[2] Ballantyne, A., et al., "Lead Acid Battery Recycling for the Twenty-First Century," *Royal Society Open Science*, Vol. 5, Issue 5, 2018, p. 171368.

[3] "Recycle Spent Batteries," *Nature Energy*, Vol. 4, No. 4, 2019, p. 253.

[4] Matsumura, N., "Cathode Technologies for Consumer Electronics," *Cathodes 2017*, California, 2017, pp. 5–6.

[5] Aguiló-Aguayo, N., et al., "Water-Based Slurries for High-Energy $LiFePO_4$ Batteries Using Embroidered Current Collectors," *Scientific Reports*, Vol. 10, 2020, p. 5565.

[6] Schreiner, D., et al, "Comparative Evaluation of LMR-NCM and NCA Cathode Active Materials in Multilayer Lithium-Ion Pouch Cells: Part I. Production, Electrode Characterization, and Formation," *Journal of the Electrochemical Society*, Vol. 168, 2021, p. 030507.

[7] Li, J., et al, "Study of the Failure Mechanisms of $LiNi_{0.8}Mn_{0.1}Co_{0.1}O_2$ Cathode Material for Lithium Ion Batteries," *Journal of the Electrochemical Society*, Vol. 162, 2015, p. A1401.

[8] Placke, T., et al., "Lithium Ion, Lithium Metal, and Alternative Rechargeable Battery Technologies: The Odyssey for High Energy Density," *Journal of Solid State Electrochemistry,* Vol. 21, 2017, pp. 1939–1964.

[9] Nishi, Y., "Lithium Ion Secondary Batteries; Past 10 Years and the Future," *Journal of Power Sources*, Vol. 100, No. 1–2, 2001, pp. 101–106.

[10] Vadlamani, B., "An In-Situ Electrochemical Cell for Neutron Diffraction Studies of Phase Transitions in Small Volume Electrodes of Li-Ion Batteries," *Journal of the Electrochemical Society*, Vol. 161, 2014, pp. A1731–A1741.

[11] Gu, M., et al., "Nanoscale Silicon as Anode for Li-ion Batteries: The Fundamentals, Promises, and Challenges," *Nano Energy*, Vol. 17, 2015, pp. 366–383.

[12] Zhu, G., et al., "Dimethylacrylamide, a Novel Electrolyte Additive, Can Improve the Electrochemical Performances of Silicon Anodes in Lithium-ion Batteries," *RSC Advances*, Vol. 9, 2019, pp. 435–443.

[13] Cho M., et al., "Anomalous Si-based Composite Anode Design by Densification and Coating Strategies for Practical Applications in Li-ion Batteries," *Composites Part B: Engineering*, Vol. 215, 2021, p. 108799.

[14] Xu, W., et al., "Lithium Metal Anodes for Rechargeable Batteries," *Energy & Environmental Science,* Vol. 7, 2014, pp. 513–537.

[15] Zhou, L., et al., "Lithium Sulfide as Cathode Materials for Lithium-Ion Batteries: Advances and Challenges," *Journal of Chemistry*, Vol. 2020, 2020, Article ID 6904517.

[16] Bruce, P., et al., "Li-$O_2$ and Li-S Batteries with High Energy Storage," *Nature Materials*, Vol. 11, No. 1, 2012, pp. 19–29.

[17] Lide, D., *CRC Handbook of Chemistry and Physics,* 86th Edition, 2005–2006, Boca Raton, FL: CRC Press, Taylor & Francis, 2005, pp. 4–70.

[18] Julien, C., "Lithium Iron Phosphate: Olivine Material for High Power Li-Ion Batteries," *Research & Development in Material Science*, Vol. 2, 2017.

[19] Manthiram, A., "A Reflection on Lithium-ion Battery Cathode Chemistry," *Nature Communications,* Vol. 11, No. 1, 2020, p. 1550.

[20] Murdock, B., et al., "A Perspective on the Sustainability of Cathode Materials Used in Lithium-Ion Batteries," *Advanced Energy Materials,* Vol. 11, Issue 39, 2021, p. 2102028.

[21] Nam., G., et al., "Capacity Fading of Ni-Rich NCA Cathodes: Effect of Microcracking Extent," *ACS Energy Letters,* Vol. 4, No. 12, 2019, pp. 2995–3001.

[22] Barkholtz, H., et al., "Multi-Scale Thermal Stability Study of Commercial Lithium-ion Batteries as a Function of Cathode Chemistry and State-of-Charge," *Journal of Power Sources,* Vol. 435, 2019, p. 226777.

[23] Beaulieu, L. Y., et al., "Colossal Reversible Volume Changes in Lithium Alloys," *Electrochemical and Solid-State Letters,* Vol. 4, No. 9, 2001, p. A137.

[24] Zhao, X., et al., "Challenges and Prospects of Nanosized Silicon Anodes in Lithium-ion Batteries," *Nanotechnology,* Vol. 32, No. 4, 2020, p. 042002.

[25] Sun, Y., et al., "Promises and Challenges of Nanomaterials for Lithium-based Rechargeable Batteries," *Nature Energy,* Vol. 1, No. 7, 2016, p. 16071.

[26] Bates, J. B., et al., "Thin-film Rechargeable Lithium Batteries," *Journal of Power Sources,* Vol. 54, No. 1, 1995, pp. 58–62.

[27] Kamaya, N., et al., "A Lithium Superionic Conductor," *Nature Materials,* Vol. 10, No. 9, 2011, pp. 682–686.

[28] Nemeth, T., et al., "Lithium Titanate Oxide Battery Cells for High-Power Automotive Applications—Electro-Thermal Properties, Aging Behavior and Cost Considerations," *Journal of Energy Storage,* Vol. 31, 2020, p. 101656.

# 6

# LI-ION BATTERY CELL/PACK DESIGN AND MANUFACTURING/RECYCLING PROCESS

## 6.1 INSIDE A LI-ION BATTERY

### 6.1.1 Battery Cell and Pack

When you see a Li-ion battery attached to a system (e.g., smartphone), it is typically a battery pack. A battery pack consists of a battery cell and a protection circuit as shown in Figure 6.1.

A Li-ion battery cell is an energy storage component with high-energy density. While it provides long battery life to a system, it also brings a safety risk when the battery is inappropriately used, such as overcharging. To mitigate the risk, a battery cell is assembled with a protection circuit. The assembly is called a battery pack. A protection circuit typically includes a fuel gauge IC (or a gas gauge IC) that monitors battery status (e.g., SOC, capacity, voltage) and offers primary protection functions, a second protection IC, and thermistor.

### 6.1.2 Cell Form Factors

For a battery cell, there are mainly three form factors, cylindrical, prismatic, and pouch, as shown in Table 6.1.

All form factors include cathode, anode, separator, and electrolyte. Cathode, anode, and separator are similar across the form factors.

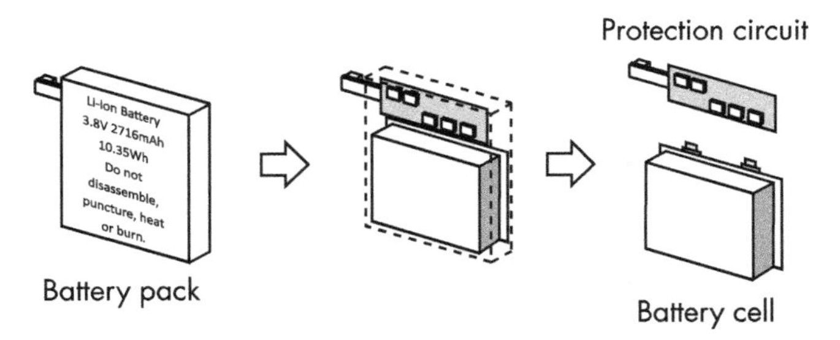


**Figure 6.1** Schematic illustration of a battery pack that consists of a battery cell and a protection circuit.

**Table 6.1**
Comparison of Cylindrical, Prismatic, and Pouch Form Factors

| **Cell Form Factor** | **Cylindrical** | **Prismatic** | **Pouch** |
|---|---|---|---|
| **Cathode (+)** | Typical cathode (e.g., $LiCoO_2$, NMC) | | |
| **Electrolyte** | Organic liquid (or polymer in pouch) | | |
| **Separator** | Typical separator (e.g., PP/PE/PP trilayer) | | |
| **Anode (–)** | Typical anode (e.g., graphite) | | |
| **Case** | Stainless steel can | Aluminum can | Aluminum sheet with insulation layers |
| **Dimensions** | Standardized (e.g., 18650, 2170, 4680) | Semi-standardized (e.g., 406080) | Customizable |
| **Advantage** | Fast manufacturing speed | Shape options with a hard case | Custom shape, thinness |

Typical cathode (e.g., $LiCoO_2$, NMC), anode (e.g., graphite), and separator (e.g., Polypropylene (PP)/Polyethylene (PE)/Polypropylene (PP) trilayer) can be used in any form factors. Organic liquid electrolyte can also be used in any form factors. In a pouch form factor, polymer electrolyte is also an option. The key difference among the form factors is a case.

The case of a cylindrical form factor is made of stainless steel. The dimensions are standardized such as 18650, 2170, and 4680 format: 18650 means ~18 mm in diameter and ~65.0 mm in length. The first two digits describe the diameter, and the last three digits describe the length. Recently 2170 and 4680 formats were introduced: 2170 means ~21 mm in diameter and ~70 mm in length. The detailed dimensions may be different from the format name. Therefore, it is important to check the specification of the cell.

***Exercise***
What are the dimensions of a 4680 cylindrical cell?

***Answer***
~46 mm in diameter and ~80 mm in length.

* * *

Because of the standardized form factor, battery manufacturing speed can be fast, resulting in low battery cost. Therefore, this form factor is popular in electric vehicles where battery cost reduction is desired.

For a prismatic form factor, the case is an aluminum can that is formed from an aluminum sheet by deep drawing. One example is a 406080 format. For prismatic and pouch form factors, the first two digits represent thickness, the next two digits represent width, and the last two digits represent length. The 406080 format is ~4.0 mm-thick × ~60 mm-wide × ~80 mm-long. Sometimes, letters are used in the format name, such as 4060A2. The letters typically correspond to two digits from 10. For example, A is 10, B is 11, and C is 12. In this case, the length of a 4060A2 format is ~102 mm. Again, the detailed dimensions may be different from the format name. Therefore, it is always important to check the specification of the cell. Compared to cylindrical cells, there are more shape options for prismatic cells. As the case is made by deep drawing, very thin and long cases are difficult to manufacture. This form factor is popular in devices where prismatic shape is preferred to cylindrical shape, but a hard case is still needed. One example is a mobile phone where the rectangular battery fits the product design, and the battery is replaceable by users.

For a pouch form factor, the case is made of an aluminum sheet laminated with insulation layers. The assembly of cathode, separator, and anode are wrapped with the sheet. Then, the edges of the sheet are heat-sealed. The pouch cell can be thin when the assembly of cathode, separator, and anode is thin. Also, the pouch cell can be nonrectangular when the cathode, separator, and anode are cut into a nonrectangular shape, assembled in a safe manner without short circuit, and wrapped with a nonrectangular pouch. An example of a nonrectangular cell is shown in Figure 6.2.

Because a pouch form factor can be thin and the shape is customizable, it is popular in the devices that need a thin and/or custom battery. For example, smartphones or laptop PCs have stylish industrial designs where a thin battery is required, and battery dimensions are different from model to model and generation to generation. The pouch form factor can meet those requirements.

In general, when a cell is thicker, volumetric energy density (Wh/l) is higher even with the same cathode and anode technologies. The thickness of the case material (e.g., pouch sheet) does not change even when the cell is thicker. However, the total thickness of the electrodes that store energy increases. As a result, volumetric energy density increases. This is also the case for width or length. Therefore, when volumetric energy densities are compared, it is important to use the same cell dimensions.

### 6.1.3 Battery Cell Structure

In a battery cell, there are cathode electrode(s), separators, and anode electrode(s). These sheets are stacked, wound, or folded, and inserted in a package with an electrolyte. Figure 6.3 is an example of Li-ion cell structure.

Figure 6.4 shows a cross-sectional illustration of cathode electrode, separator, and anode electrode.

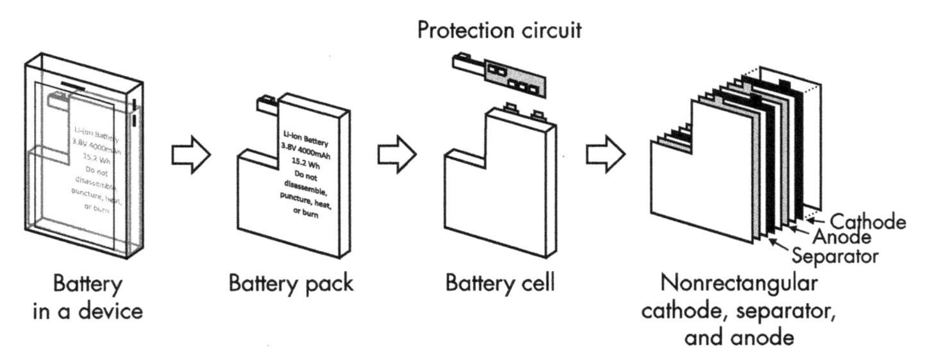


**Figure 6.2** An example of a nonrectangular pouch cell.

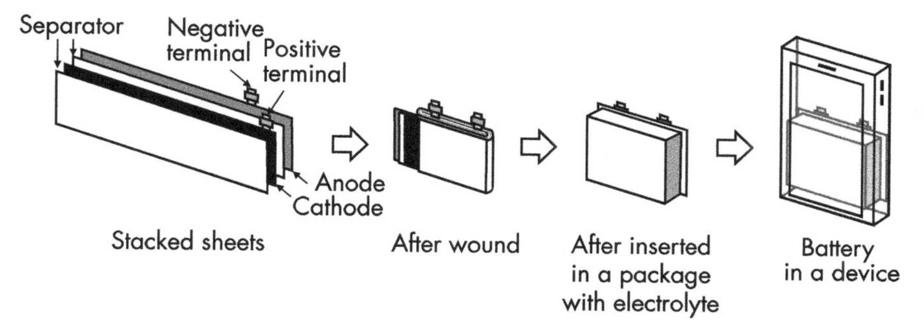


**Figure 6.3** An example of Li-ion cell structure.

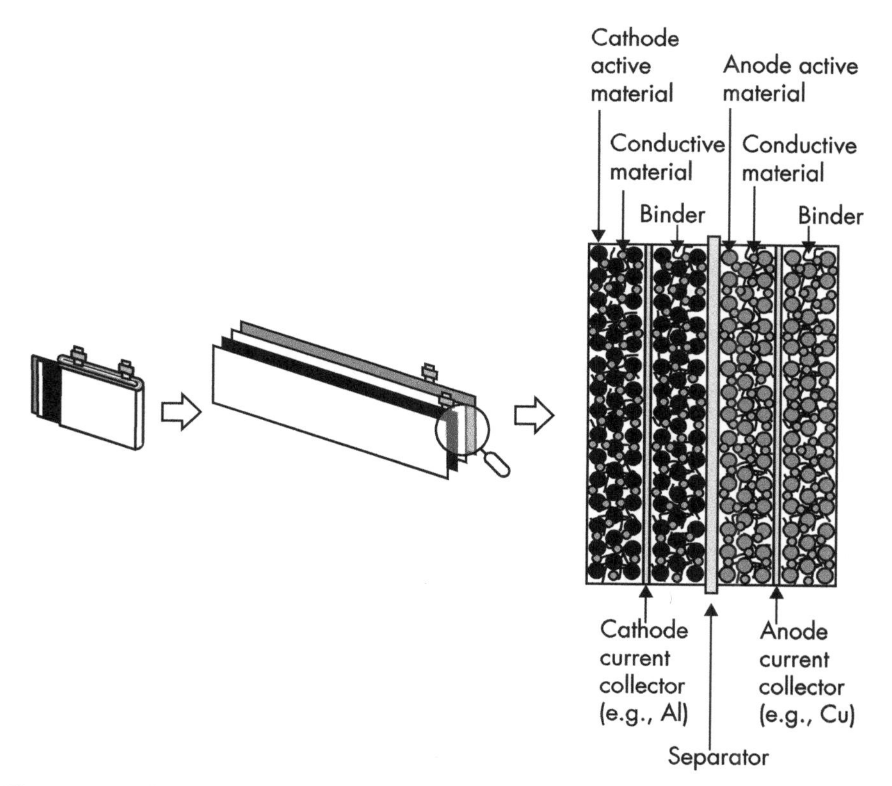


**Figure 6.4** A cross-sectional illustration of cathode electrode, separator, and anode electrode.

For the cathode electrode, the mixture of cathode active material (e.g., LCO), conductive material (e.g., carbon), and binder (e.g., polyvinylidene difluoride, which is called PVDF) is pasted on a current collector, which is a metal sheet (typically aluminum) [1, 2]. The particles of the cathode active material have a large surface area where lithium ions can move from/into the cathode active material at reduced impedance. The particles of the conductive material help electrons move from/into the active material and reduce impedance. The binder helps contact between cathode active material, conductive material, and current collector. It also maintains the reduced impedance. The material of the metal sheet needs to be electrochemically stable under the voltage range of Li-ion battery charging and discharging, such as 3.0V to 4.4V. Aluminum is typically chosen for a cathode current collector because thin aluminum oxide formed on the surface of the aluminum sheet works as a protective film and enables stability over the voltage range. Both sides of the current collector are coated with the cathode active material with conductive material and binder. With the double-sided coating, cathode and anode are always facing each other through the separator after being wound, folded, or stacked. If it is a single-sided coating, there are some areas where the cathode and anode current collectors face each other without active materials, which is not efficient from an energy density standpoint.

For the anode electrode, the mixture of anode active material (e.g., graphite), conductive material (e.g., carbon), and binder (e.g., styrene-butadiene rubber, which is called SBR with carboxymethyl cellulose, or CMC) is pasted on a current collector that is a copper sheet [1, 2]. The roles of conductive material and binder are the same as in the cathode electrode, such as impedance reduction. The anode current collector is copper because it is stable under the voltage range of the battery operation and has high electrical conductivity. Similar to the cathode electrode, both sides of anode current collector are coated with active material, conductive material, and binder to provide high-capacity efficiently.

For a Li-ion battery, a nonaqueous electrolyte, such as carbonate, is used because of its stability in the battery voltage range. During charge and discharge, lithium ions need to move between cathode and anode through the electrolyte. To support this, a lithium salt such as lithium hexafluorophosphate ($LiPF_6$) is dissolved in the electrolyte so that the electrolyte contains lithium ions. Electrolyte also includes additives to enhance low- or high-temperature performance, cycle life, and safety. These additives are often the "secret sauce" of the suppliers to enable new performance capabilities.

The separator is a polyolefin film, such as PP/PE/PP trilayer. It is electrically insulative and porous. The electrolyte fills the pores and lithium ions can move through the separator during charge and discharge. When abnormal current flows and the generated heat increases above a certain level, part of the separator typically melts and closes the pore. This stops ion movement and current flow, mitigating a risk of a hazardous situation. The separator is required to have sufficient mechanical strength and thermal stability. For example, it should not shrink at elevated temperatures. If it shrinks, cathode and anode touch each other, and a short circuit happens. There are also functional separators where the surface is coated with thin layers of ceramic, PVDF, and so forth. This additional layer provides enhanced safety against heat and short circuit.

Cases/pouches of the battery are already explained in the earlier section. One important fact is that the side of cylindrical and prismatic cases is also part of the negative terminal. Usually, the side is wrapped with an insulative film for safety as shown in Figure 6.5. However, if a battery cell holder is made of metal and scratches the film after battery insertion, an unexpected external short circuit may happen.

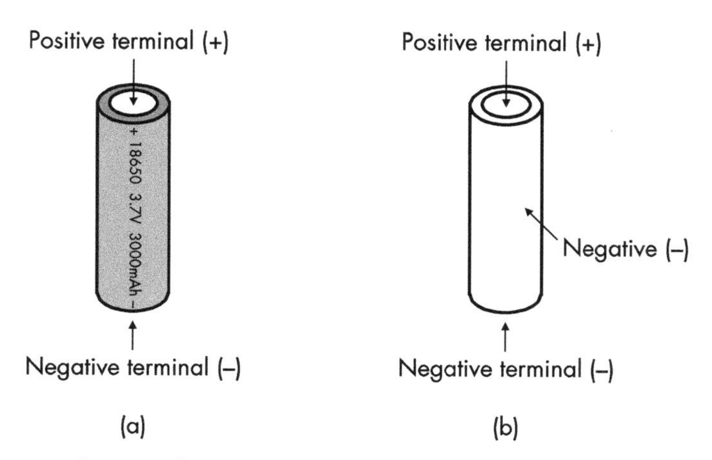


**Figure 6.5** Schematic illustration of positive and negative terminals for a cylindrical cell (a) with and (b) without a film on the side.

### 6.1.4 Cell Manufacturing Process

Figure 6.6 shows an example of a cell manufacturing process.

First, cathode active material, conductive material, and binder are mixed in a nonaqueous solvent, such as NMP (N-Methyl-2-pyrrolidone). The mixture is called slurry. Similarly, anode slurry is made by mixing anode active material, conductive material, and binder in water. Cathode needs nonaqueous solvent because water reacts with cathode material and decreases the cell capacity.

Then, both sides of current collectors are coated with the slurry. The cathode current collector, which is typically aluminum, is coated with the cathode slurry. The anode current collector, which is typically copper, is coated with the anode slurry. Each electrode goes through an oven to dry the solvent in the slurry. Since NMP is toxic and requires more energy to dry than water, an aqueous process or dry process is desired and has been researched for cathode.

After drying, the electrode goes through the calendering process where the two rollers compress the electrode. The calendering process increases the cell energy density and helps active materials adhere to the current collectors. The pressure used to calender helps determine the final density and porosity of the electrode, which in part is driven by the desired energy density and rate capability of the cell.

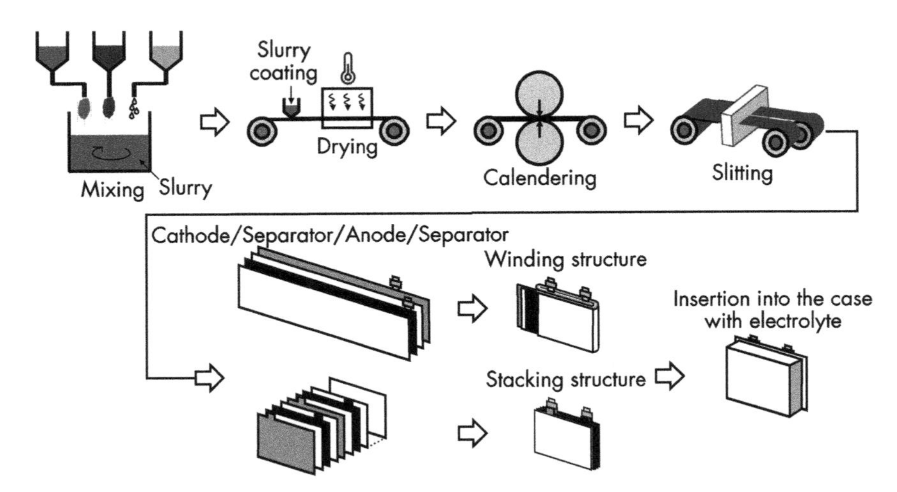


**Figure 6.6** An example of a cell manufacturing process.

After cathode and anode electrodes are made, the electrodes and separator are slit into the strips with the desired width, or cut into the rectangular pieces with the desired width and length. Burrs during slitting/cutting need to be prevented to avoid a short circuit after assembly.

After tabs, which are the positive/negative terminals, are attached to the cathode and anode, the strips of cathode, separator, anode, and separator are wound together to form the cell assembly with the winding structure. In the case of the rectangular pieces of cathode, separator, anode, and separator, these are stacked to form the cell assembly with the stacking structure. Careful alignment between cathode, separator, and anode is required to avoid a short circuit.

The assembly is inserted into the case with electrolyte. Figure 6.6 shows an example of a pouch cell. The winding structure can be used for cylindrical, prismatic, and pouch form factors. The stacking structure is only for pouch form factors because it is difficult to insert the structure into the cylindrical or prismatic case.

After electrolyte is injected, the reactions between electrolyte and active materials start and form SEI on the surface of the active materials. During initial charging and discharging cycles, additional SEI formation happens. While SEI formation consumes precious lithium, the SEI prevents further reaction between active materials and electrolyte, resulting in prevention of lithium consumption and better cycle life. Therefore, a uniform SEI needs to be formed. The formation and aging process is part of the manufacturing process after electrolyte injection. This is to form a uniform SEI layer. As a result of the reactions, gas is also generated. As the gas takes up space and decreases energy density, it is removed from the cell. For example, in case of a pouch cell, there is a pouch pocket that is a redundant pouch area for the gas to build up. During manufacturing, pressure is applied to the cell to expel the gas to the pocket. Then, the pocket is cut out of the cell while the edge of the cell is heat-sealed to form the final pouch cell. The electrolyte needs to be treated in a dry environment and the moisture from the electrodes needs to be removed in advance. This is because $LiPF_6$ in the electrolyte reacts with moisture ($H_2O$) and forms hydrogen fluoride (HF), which is highly acidic and corrosive.

In any process, metallic contaminants need to be avoided. They may cause an immediate short circuit after assembly. They may also be electrochemically dissolved into electrolyte and plated on electrodes over discharging/charging cycles, which results in an internal short circuit.

In general, the process of stacking structure is more difficult than that of winding structure because more careful alignment of electrodes and separators is needed to avoid internal short circuits. However, it provides lower impedance because the electron path to the terminals is shorter than that in the winding structure, which has a long strip and long distance to the terminals. Stacking structure also enables a nonrectangular cell, such as an L-shaped cell, as shown in Figure 6.2.

On a side note, there are some approaches being explored to replace the electrolyte with an SSE as explained in Chapter 5, leveraging the existing stacking-up process of Li-ion batteries. This is because all-solid-state batteries with an SSE have a potential to enable higher energy density with better safety than today's Li-ion batteries with a flammable electrolyte. For example, the composite of a cathode-active material and an SSE is coated on a cathode current collector as a cathode electrode, an SSE sheet is used instead of a separator without injection of a conventional electrolyte, and a lithium metal anode is used instead of graphite. These are stacked up and high pressure is applied. While lithium metal anode contributes to high-energy density, it is challenging to maintain good contact between the SSE and the electrodes during cycles, resulting in impedance increase and worse cycle life [3]. Continuous application of high pressure to the battery during operation may mitigate the challenge. However, that may require a thick enclosure. If an enclosure with several millimeter thickness is used for a several-millimeter-thick battery, total energy density is substantially decreased. Therefore, large batteries such as electric vehicle batteries may benefit from the technology because the impact of enclosure thickness is small. Utilizing an existing chassis as part of the enclosure may also be an option for better space utilization. Further research and development are expected to improve cycle life.

### 6.1.5 Thin-Film Battery Manufacturing Process

Some micro devices (smart cards, sensors, etc.) need thin batteries. Such batteries may be manufactured through the process explained in the previous section. However, when a very thin battery, such as a several tens of $\mu$m-thick battery, is needed, a thin-film all-solid-state battery is one option to power such devices. Figure 6.7 is a cross-sectional image for an example of a thin-film all-solid-state battery [4].

The thin-film all-solid-state battery is fabricated with the vacuum process. First, current collectors are deposited on a substrate by sputtering. For example, platinum current collectors are deposited on a glass or aluminum oxide substrate. Then, cathode material is deposited on a cathode current collector. As the deposited cathode material is amorphous, the entire cell is annealed to form the layered crystal structure in the cathode. Solid-state electrolyte (e.g., LiPON) is sputtered to cover the cathode. After that, anode material (e.g., lithium) is deposited by evaporation to face the cathode through the solid-state electrolyte. As all materials are sensitive to moisture, the cell is coated with a protective layer.

For more on solid-state batteries, which are being explored for applications beyond thin batteries, all the way up to electric vehicles, see the previous section and Chapter 5.

## 6.2 PREVENTION OF HAZARDOUS SITUATIONS

### 6.2.1 Hazardous Situations

A Li-ion battery has high-energy density. Misuse of the battery brings a risk of hazardous situations such as the following:

*Overvoltage:* When a battery is charged to higher voltage than the spec, too much lithium is extracted from the cathode material and goes to anode. Because such an amount of lithium is more than the capacity the anode can accept, the excess lithium is deposited on the anode as lithium metal. As lithium metal may grow as a dendrite, there is a risk of internal short circuits. Also, overvoltage is outside the safe voltage window for electrolyte and cathode, which causes

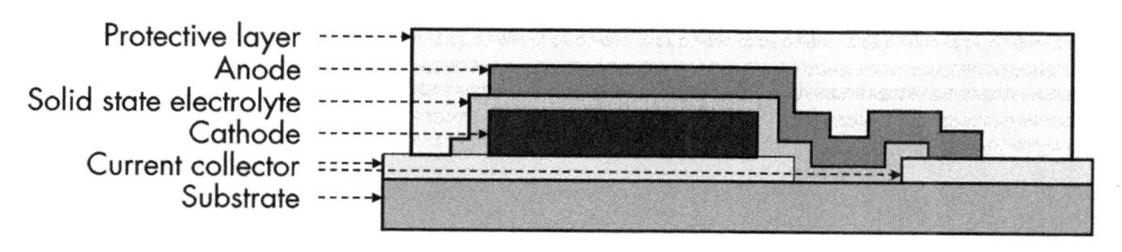


**Figure 6.7** A cross-sectional image for an example of a thin-film all-solid-state battery.

decomposition of electrolyte and collapse of the cathode crystal structure, resulting in an explosion risk and worse cycle life.

*Undervoltage:* When a battery is deeply discharged and the voltage is below safe cutoff voltage, copper current collector dissolves and SEI is decomposed, resulting in worse cycle life and a safety risk [5].

*Overcurrent:* When the charging or discharging current exceeds the spec, overheating happens due to joule heat by battery impedance. Such a high current also transports lithium between cathode and anode active materials at a higher rate than the lithium diffusion speed in the crystal structure of the active materials, resulting in local overcharging/overdischarging.

*Over-/undertemperature:* When battery temperature is high either due to the ambient temperature or overheating as a result of overcurrent, degradation reactions, such as electrolyte decomposition, are accelerated [6–8]. When battery temperature is low, ionic mobility slows down in a battery cell, and lithium ions may be plated on the surface of anode during charge instead of being safely stored in the anode material. If lithium plating happens, dendrites may grow and cause an internal short circuit.

*Short circuit:* In addition to the reasons above, short circuits may happen due to electrode deformation, misalignment, and metal contamination. Usually, a separator prevents cathode and anode electrodes from directly contacting each other. However, when electrodes deform after charging and discharging, the edges of cathode and anode may touch each other, resulting in a short circuit. The same situation may happen when cathode, separator, and anode are misaligned during manufacturing. When metal pieces are contaminated in the cell manufacturing, they may pierce a separator. Even if each piece is smaller than the thickness of the separator, after charging and discharging, the metal pieces may be electrochemically dissolved and then plated as a dendrite, causing an internal short circuit. Metal contamination is also a risk when a battery pack is assembled because it may cause an external short circuit.

### 6.2.2 Battery Swelling

Pouch or prismatic cells swell after charging, cycles, or storage.

Swelling after charging is because of the structural change of cathode and anode. For example, in a Li-ion battery with $LiCoO_2$

cathode and graphite anode ($C_6$), cathode changes to $Li_{1-x}CoO_2$ and anode changes to $LiC_6$ after charging. Such changes affect the thickness of electrodes and the cell (e.g., 3% swelling in cell thickness after charging). This type of swelling is reversible, which means that the battery shrinks close to the original thickness after discharging.

However, swelling after many charging/discharging cycles or long-term storage is irreversible. Depending on the history of battery usage, there are several reasons for the irreversible swelling, such as growth of resistive layers (SEI) due to parasitic reactions over cycles, and gas generation due to electrolyte decomposition [7, 8].

Figure 6.8 is an example of abnormal Li-ion battery swelling due to gas generation.

In a system, clearance is typically allocated between a new battery and the adjacent chassis/component to accommodate battery swelling after cycles/storage. One example is 8% swelling in battery thickness after 500 charging/discharging cycles at room temperature. However, when a battery is exposed to high temperature, worse swelling may happen. For example, when a battery is stored at high temperature or is charged/discharged fast, electrolyte decomposition is accelerated, resulting in more and faster swelling, and also high SOC, which means high cell voltage, accelerates swelling. Battery space in a system (e.g., smartphone) needs to be designed to sufficiently accommodate such swelling. Otherwise, the swollen cell is pushed back by the system (e.g., chassis) and hazardous situations, such as battery internal shorts, may happen.

### 6.2.3 Safety Protections from Failure Modes

To mitigate the safety risks, multiple safety functions are typically incorporated in a cell and pack. In a battery pack, there is a protection circuit as shown in Figure 6.1. The protection circuit includes several

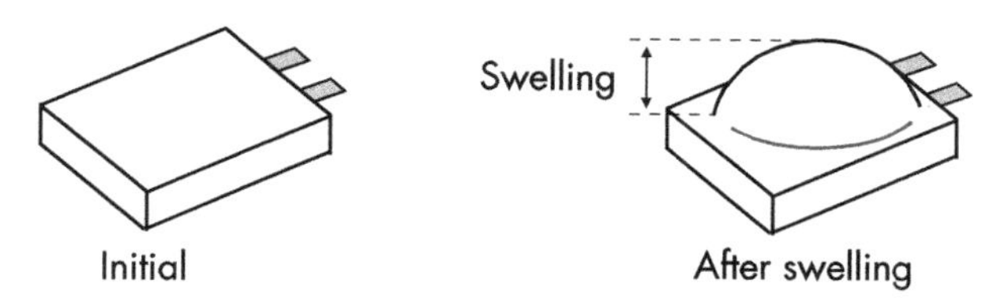


**Figure 6.8** Schematic illustration of abnormal Li-ion battery swelling due to gas generation.

protection functions and is called PCM or protection circuit board (PCB). Figure 6.9 is an example of the protection circuit.

In a battery pack, the PCM is connected to battery cell(s). Figure 6.10 is a simplified illustration of a battery pack where three cells connected in series are protected by a PCM that includes a fuel gauge IC, additional protection IC, and fuse.

In the figure, there is a fuel gauge IC that is sometimes abbreviated as FG and is also called a gas gauge IC. Some fuel gauge ICs not only gauge the remaining battery capacity, but also monitor cell/pack voltage, charging/discharging current, and battery temperature. When over-/undervoltage, overcurrent or over-/undertemperature situations happen, the IC detects it and takes actions such as lowering battery charging voltage, and/or regulating or stopping the current by software and hardware. To stop charging/discharging current, pin i or h opens one of the two field-effect transistors (FETs) that are for charging and discharging, respectively. Software of the IC offers the first-level protection and hardware offers the second-level protection. For traceability, some ICs also log the key measured values, such as maximum cell voltage, current, and temperature.

For temperature detection, some battery packs have a temperature sensor, such as an NTC thermistor. NTC stands for negative temperature coefficient. The resistance of an NTC thermistor decreases when temperature increases. The IC reads the resistance, such as through pin g in Figure 6.10, and converts it to temperature. The protection circuit in a battery pack and/or a charger IC in a system have temperature limits defined for charging and discharging. If the

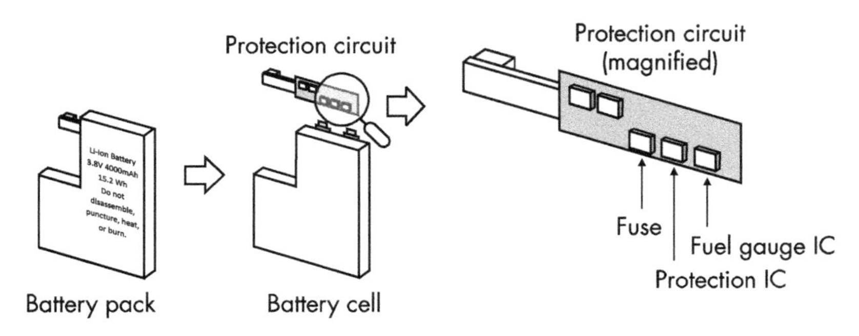


**Figure 6.9** An example of a battery protection circuit.

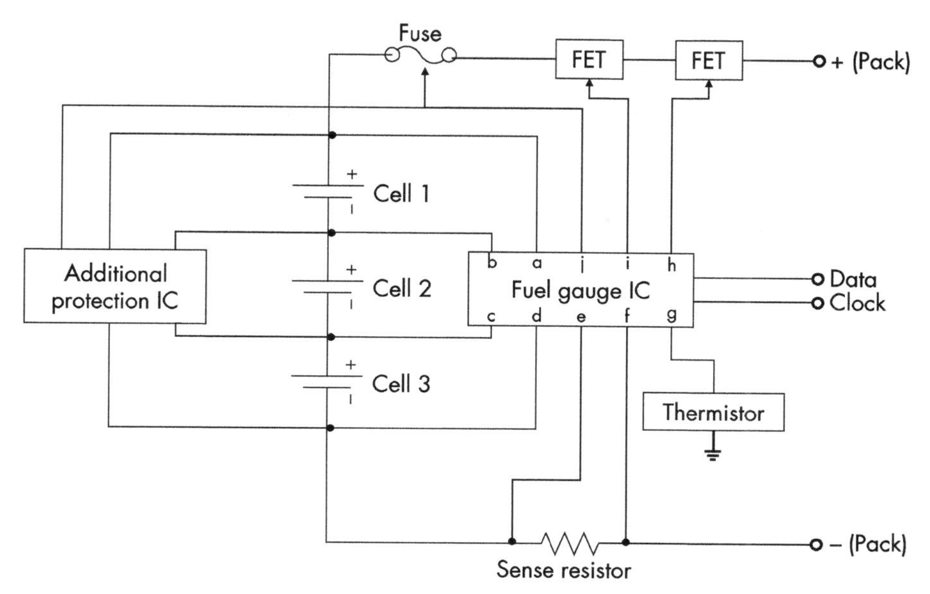


**Figure 6.10** A simplified illustration of a battery pack where three cells connected in series are protected by a PCM that includes fuel gauge IC, additional protection IC, and fuse.

temperature sensor detects temperatures outside these limits, that operation is either disallowed or scaled back.

A fuel gauge IC reports the battery status to the system through line(s) such as data and clock lines in Figure 6.10. The communication protocol is explained in Section 6.3.5.

In addition to the fuel gauge IC, there may be an additional protection IC in a protection circuit as shown in Figure 6.10. It typically protects a battery from overvoltage, undervoltage, and overcurrent.

Some battery protection circuits have a fuse in a current path. In an abnormal situation, the fuse is blown, for example, by a fuel gauge IC through pin j in Figure 6.10. This permanently disables the battery pack from charging or discharging. There is another type of fuse that has positive temperature coefficient (PTC). When high current flows through the PTC fuse and its temperature exceeds a certain level, its resistance substantially increases. This temporarily limits current flow until the resistance decreases with temperature decrease. While a fuse provides better safety, additional impedance to the current path, space, and cost need to be considered.

In addition to the protections in a PCM, a battery cell has some safety protections. For example, some separators have a shut-down function where part of the separator melts when the temperature is above a certain level, and stops ion movement and current flow. Some cells also have a TCO device in the terminal. TCO stands for thermal cut-off. It opens the circuit when heated. Similar to a fuse, additional impedance, space, and cost need to be considered to use a TCO.

Battery charging IC also offers battery protection functions. For example, there is a timer function that checks if battery voltage increases to a certain level within a certain time during precharge. This is to avoid charging a battery that may have an internal short circuit. Overvoltage protection stops charging when battery voltage is above the limit. It also stops charging when input voltage or current from a power adapter (e.g., an AC adapter) is higher than a certain level.

To ensure safe operation of a system with a Li-ion battery, there are several safety regulations and certifications that are explained in Section 6.3.3.

### 6.2.4 Quality Inspections

Through battery supply chains, battery quality is checked many times. Figure 6.11 is an example.

Battery cell manufacturers check the quality of incoming materials, such as cathode/anode active materials, conductive materials,

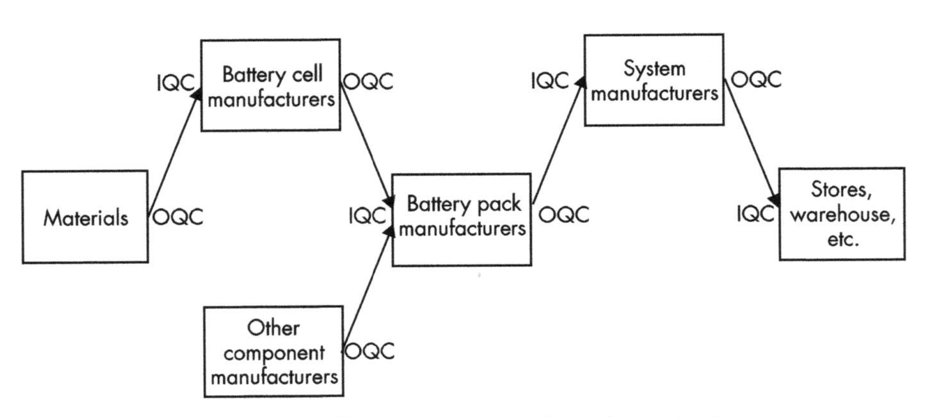


**Figure 6.11** An example of battery inspections through supply chain.

separator, and electrolyte. This is called incoming quality check (IQC). When battery cells are made, the quality is checked as outgoing quality check (OQC). The OQC includes the key items in the spec, such as capacity, impedance, dimensions, and appearance. The appearance check is to make sure that the label is clearly printed and there are no wrinkles in pouch cells, among other issues. Cycle life is also checked with some samples in a batch. The cells are delivered to a battery pack supplier that assembles the cells with protection circuits. It also puts the assembly in an enclosure or wraps it with a mylar, which is a film. To make protection circuits and battery packs, battery-pack suppliers receive cells and other necessary components, such as fuel gauge ICs, protection ICs, connectors, and mylars. Physical or document checks are conducted for these as part of IQC. After battery packs are made and their quality is checked as OQC, the packs are sent to system manufacturers, such as smartphone manufacturing companies. Sometimes, a system seller, which puts their company brand to the system, contracts with system manufacturing companies and outsources manufacturing to the companies. Therefore, such system manufacturing companies are called contract manufacturers (CM) or electronics manufacturing services (EMS). The CM performs IQC for the incoming battery packs and OQC for the systems before going to the stores and warehouse. The quality and safety of batteries are ensured through the repeated IQCs and OQCs in the supply chain.

If you design or use a battery in a project, key specifications need to be inspected somewhere as part of IQC/OQC. Table 6.2 is an example of a battery pack OQC.

Throughout the process, quality engineers may also use sampling inspections with differently defined acceptance quality limit (AQL) depending on the criticality of the defect.

### 6.2.5 Safe Battery Tests

To simulate the longevity of a battery in a product, accelerated testing is performed in the laboratory by cycling the battery repeatedly between empty to full SOC. In the cycle test, a battery testing system is used, and the battery capacity is checked at every charging and discharging cycle. The system is called a battery tester or a battery cycler. The tester typically comes with a PC where the tester control software runs. Users can easily set up the test procedure with the software.

**Table 6.2**
An Example of a Battery Pack OQC

| Inspection Item | Test Method | Sampling Size | Success Criteria |
|---|---|---|---|
| Capacity | Charge: 0.5C CC-CV to 4.35V (0.05C cutoff)<br>Discharge: 0.5C discharge to 3.0V | 100% | Minimum 4.0 Ah as discharged capacity |
| Cycle life | Charge: 0.5C CC-CV to 4.35V (0.05C cutoff)<br>Discharge: 0.5C discharge to 3.0V<br>Repeat charging/discharging for 500 cycles. | Sampling (3 samples every batch) | At least 80% of initial capacity after 500 cycles |
| Protection functions | Overcharge voltage, overcharge current, overdischarge current | 100% | Open circuit |
| Impedance | 1-kHz AC impedance at 50% SOC at room temperature | 100% | Maximum 100 mohm |
| Shipping SOC | Tester | 100% | Not to exceed 30% |
| Battery size | Go/no-go jig | 100% | Maximum 4.0 × 100.0 × 200.0 mm |

Table 6.3 is an example of a 500-cycle test program in the software interface.

In this example, charging starts with CC at 0.5C to 4.35V (step 1). When voltage reaches 4.35V, CV charging at 4.35V follows (step 2). Charging completes when current is reduced to 0.05C. Data is recorded at every 10 seconds in steps 1 and 2. After charging, 10 minutes rest time follows in step 3. This is to relax the battery before the next

**Table 6.3**
An Example of a 500-Cycle Test Program

| Step | Type | Mode | Value | End Condition | Recording Interval |
|---|---|---|---|---|---|
| 1 | Charge | CC | 0.5C | Voltage ≥ 4.35V | 10 sec |
| 2 | Charge | CV | 4.35V | Current ≤ 0.05C | 10 sec |
| 3 | Rest | — | — | 10 min | 10 sec |
| 4 | Discharge | CC | 0.5C | Voltage ≤ 3.0V | 10 sec |
| 5 | Rest | — | — | 10 min | 10 sec |
| 6 | Repeat between step 1 and 5 | — | — | 500 cycles | — |

step. In step 4, the battery is discharged at 0.5C to 3.0V with a data recording interval every 10 seconds. After discharging, 10 minutes rest time follows in step 5. In step 6, the test goes back to step 1 to repeat between step 1 and step 5 for 500 cycles.

Figure 6.12 shows the first few cycles in the cycle test where the cycle test program in Table 6.3 is applied to a 3.3-Ah cell.

Figure 6.12(a) shows how battery voltage and current change over test time. At the end of each cycle, discharged capacity is recorded and is shown in Figure 6.12(b).

While the program in Table 6.3 correctly performs the cycle test conditions, there is a potential risk. For example, if an internal short circuit happens in a cell during the cycles and current leaks within the cell, it may never complete step 1 or step 2. There are several methods to mitigate the risk. Some software allows the users to add additional end conditions in each step and such additional end conditions can

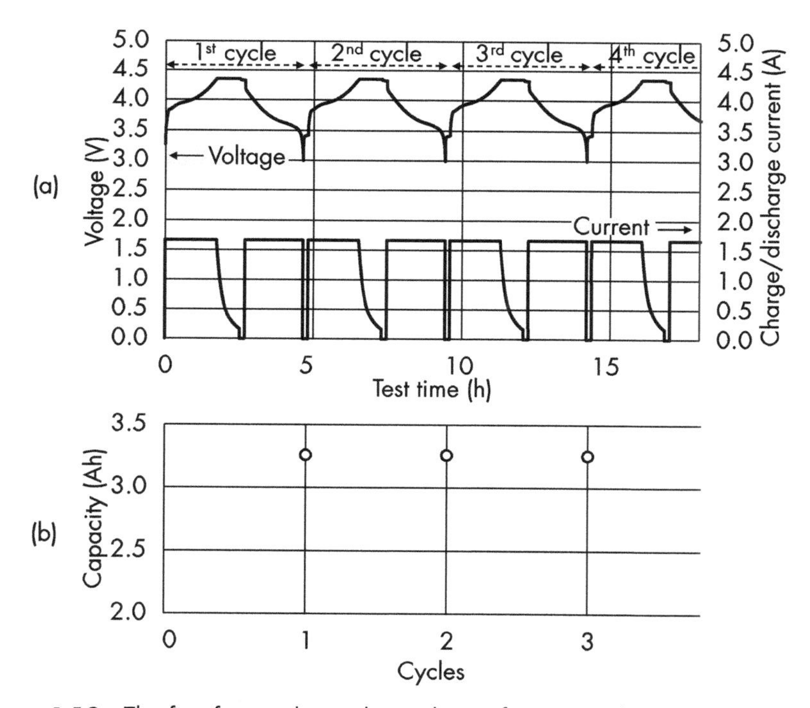


**Figure 6.12** The first few cycles in the cycle test for a 3.3-Ah cell: (a) voltage and current over test time, and (b) discharged capacity at the end of each cycle.

be configured to stop the test when the conditions are met. For example, adding time limit and/or charged capacity limit as an additional end condition in step 1 and 2 mitigates the risk. If thermocouples can be connected to the tester, the risk can also be mitigated by measuring cell temperature and using temperature limit as an additional end condition. These additional end conditions may apply to both laboratory testing and in-field system SW.

It is also important to add another end condition that stops the test when discharged capacity at the end of discharging step (step 4) is below the threshold (e.g., 80% of the initial capacity).

Some software can also set voltage limits. This mitigates a risk caused by human error. For example, for the cell where the charging voltage is 4.35V in the spec, a test operator may mistakenly write higher charging voltage (e.g., 4.65V) in a test program. If the voltage limit of the tester is set to 4.38V, the tester automatically stops the test and shows an error.

## 6.3 BATTERY PACK CONFIGURATION

### 6.3.1 Series and Parallel

Some systems have multiple connected cells in a battery pack. In the battery industry, there is a unique terminology for the connection of the cells in a battery pack. For example, some laptop PCs have a 2S2P battery pack. What does 2S2P mean? S and P explain the configuration of the battery cells in the battery pack. S means series and P means parallel. 2S2P means that two cells are connected in parallel (2P), forming a 2P block, and two 2P blocks are connected in series (2S). Figure 6.13 shows a block diagram of a 2S2P battery pack. For convenience's sake, this section shows only cells in a battery pack and does not show other components such as PCM.

When a battery pack consists of multiple S and P configurations, P blocks are usually made first, and the blocks connected in series are made later. This is to mitigate the imbalanced cells, that are explained in the next section.

***Exercise***

Write block diagrams of 2S3P, 3S2P, and 1S2P.

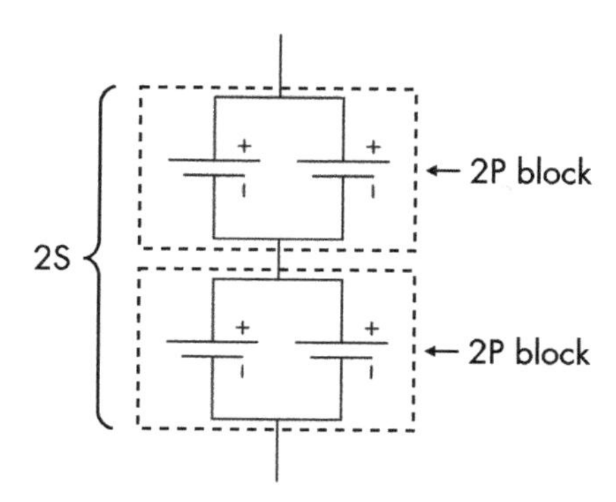


**Figure 6.13** A block diagram of a 2S2P battery pack.

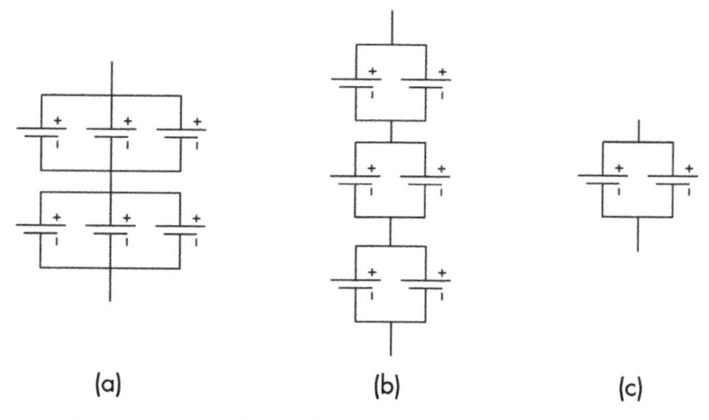


**Figure 6.14** Block diagrams of (a) 2S3P, (b) 3S2P, and (c) 1S2P.

***Answer***

Refer to Figure 6.14.

* * *

When two 1-Ah cells are connected in parallel to form a 1S2P pack and each cell is at 3.8V with 30 mohm, what is the battery pack voltage, capacity, and impedance? As two cells are connected in parallel, the battery pack voltage is 3.8V. The capacity is $2 \times 1$ Ah = 2 Ah. Impedance is 30 mohm/2 = 15 mohm, if only cell impedances are considered.

When the same two cells form a 2S1P pack, what is the voltage, capacity, and impedance? The answer is shown in Figure 6.15.

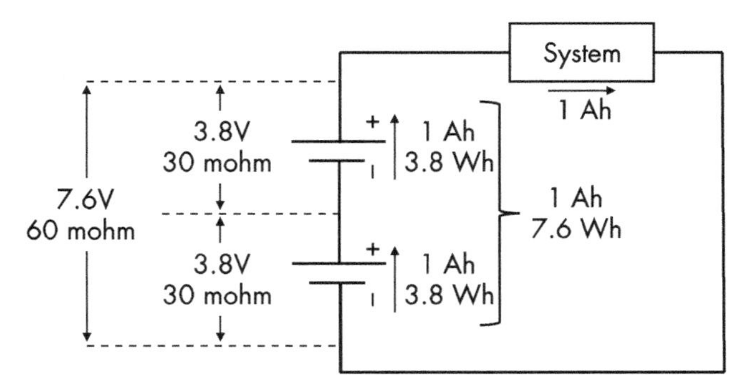


**Figure 6.15** An example of voltage, capacity, and impedance of a 2S1P battery pack.

As two cells are connected in series, voltage is doubled to 7.6V. Impedance is also doubled to 60 mohm because there are two impedances in the current path. However, the battery pack capacity is 1 Ah, not 2 Ah. This is because, when the 2S1P pack supplies 1A to a system, each cell supplies 1A and the system receives 1A. If 1A lasts for 1 hour, each cell supplies 1A for 1 hour, which is 1 Ah, and the system receives 1A for 1 hour, which is 1 Ah. As the 1 Ah comes from the battery pack, the battery pack capacity is 1 Ah. Although the pack capacity (1 Ah) is the same as the cell capacity (1 Ah), the battery pack voltage (7.6V) is twice as high as the battery cell voltage (3.8V). This means that total energy of the battery pack is 1 Ah × 7.6V = 7.6 Wh. This is equivalent to the total energy of two cells where each cell has 1 Ah × 3.8V = 3.8 Wh as shown in Figure 6.15.

The number of S series is typically determined by considering the system voltage requirement, system power, and the efficiency of voltage regulation. Figure 6.16 is an example of discharge curves for 1S and 2S batteries.

For example, when a low-power system requires at least 2.8V, as shown in case 1 of Figure 6.16, a 1S Li-ion battery with 3.0V-discharge cutoff is sufficient. This is because most of the energy in the battery is usable above the system voltage requirement. However, when a system requires at least 5.8V (case 2) and sometimes draws high current from the battery, a 2S Li-ion battery or higher S is needed. This is because battery voltage drops during discharge due to IR drop and still needs to meet the system voltage requirement. As higher discharge current causes higher IR drop, it may be safer to have a more than 2S

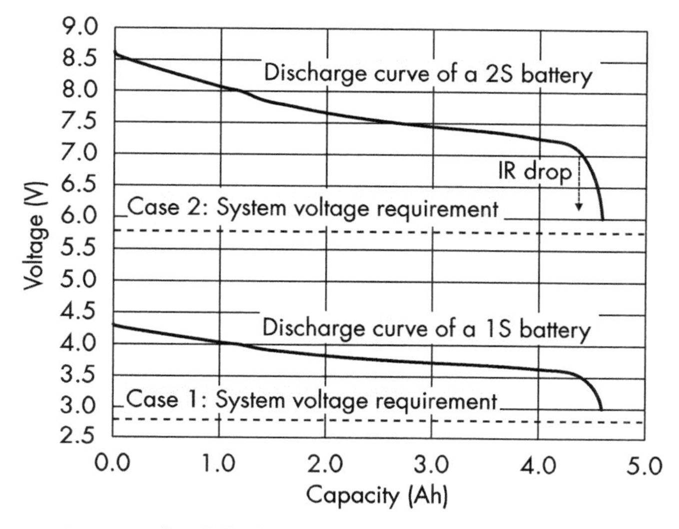


**Figure 6.16** An example of discharge curves for 1S and 2S batteries.

battery (e.g., 3S) to avoid sudden system shutdown. There is a way to use a 1S battery and boost the voltage with a step-up voltage regulator for the system in case 2. However, inefficiency in voltage conversion results in less usable energy from the battery.

On a side note, battery pack impedance consists of not only the impedances of cells, but also the impedances of a connector, a protection circuit, TCO, and so forth. These need to be considered to calculate the IR drop.

### 6.3.2 Impact of Imbalanced Cells

We already know that batteries are degraded over time and full-charge capacity decreases. When a battery pack includes several cells and the cells are the same model from the same batch, are those cells degraded at the same speed? The answer is no. Figure 6.17 is an example of battery cycle life for two cells from the same model and the same batch based on the same test condition.

As shown in the figure, there is a slight difference in the initial battery capacity. The capacity difference increases after cycles. If one cell in a battery pack is closer to a heat source, such as a processor, than the other cells, it will degrade faster, and the capacity difference will be even larger. This means that the full-charge capacity of each

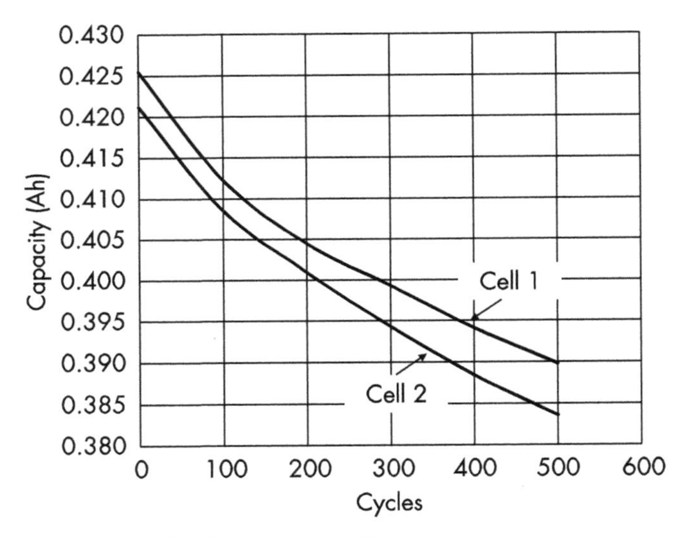


**Figure 6.17** An example of battery cycle life for two cells from the same model and the same batch based on the same test condition.

cell in a battery pack will be different over time. Such imbalanced capacities will be a problem when the cells are connected in series. Figure 6.18 shows an example of imbalanced cells in a 2S1P battery.

When there is a capacity difference between the cells connected in series, one cannot be fully charged or fully discharged. For example, cell 1 (maximum 2.0 Ah) and cell 2 (maximum 1.5 Ah) in Figure 6.18 have different full-charge capacities due to design difference or

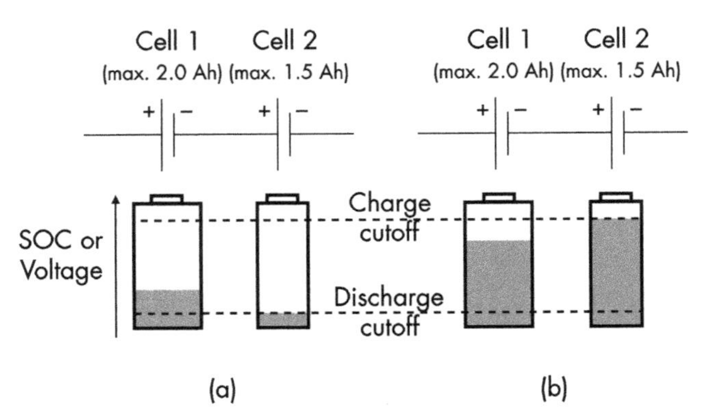


**Figure 6.18** Impact of imbalanced cells on capacity in case of a 2S1P battery (a) after discharging, and (b) after charging.

degradation difference. During discharge, both cell 1 and cell 2 supply the same current and the same Ah. Due to full-charge capacity difference, cell 2 may reach discharge cutoff voltage earlier than cell 1 and battery discharging stops. If discharging continues, cell 1 reaches discharge cutoff voltage but cell 2 is overdischarged, which brings a safety risk. During charge, both cell 1 and cell 2 receive the same current and the same Ah. Due to full-charge capacity difference, cell 2 may reach charge cutoff voltage earlier than cell 1 and battery charging stops. If charging continues, cell 2 is overcharged.

As these examples show, a battery pack with imbalanced cells connected in series cannot be fully charged or discharged, resulting in reduction of total battery capacity. Also, if the system monitors only the battery pack voltage and does not monitor each cell voltage, overcharging or overdischarging happens. When a battery pack consists of multiple S series, the capacity of each cell or cell block needs to be the same.

On a side note, some ICs or battery designs may have a cell-balancing function. This allows current to bypass some cells connected in series typically during charge, and fully charges all cells (cell-balancing). Still, there is a full-charge capacity difference between the cells connected in series even after cell balancing. The system and battery designs that mitigate the imbalanced capacity risk are desired, such as selecting the same capacity cells to make a battery pack, and keeping the battery pack away from a heat source in the system.

Then, what if two cells with different capacities are connected in parallel? Figure 6.19 is an example of different-capacity cells connected in parallel as 1S2P.

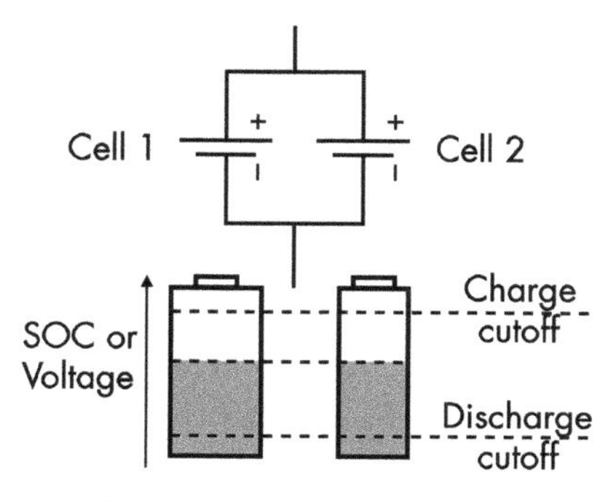


**Figure 6.19** An example of different-capacity cells connected in parallel as 1S2P.

In this case, whether capacity difference is due to design difference or degradation difference, battery voltages of cell 1 and cell 2 are the same because of the parallel connection. The current flow to or from the 1S2P battery during charge or discharge is split between cell 1 and cell 2, depending on each cell impedance and OCV. It is important to make sure that the current to each cell is within the spec of each cell.

This parallel connection enables the connection of different-sized cells that have different capacities when the charge and discharge cutoff voltages are the same. For example, a 1.5-Ah cell and a 2.0-Ah cell can be connected in parallel, which can form a nonrectangular 1S2P battery pack with 3.5 Ah as shown in Figure 6.20.

By utilizing this, it is also possible to design a nonrectangular battery pack with multiple S series. Figure 6.21 is an example of a nonrectangular 3S2P battery pack with three different-sized cells.

In this figure, three types of different-sized cells are used: 3.6 Ah, 3.8 Ah, and 4.0 Ah. Charge and discharge cutoff voltages of these cells are the same. A 3.6-Ah cell and a 4.0-Ah cell are connected in parallel, forming the 7.6-Ah block 1. Block 2 also consists of 3.6-Ah and 4.0-Ah cells connected in parallel, forming the 7.6-Ah block. Block 3 consists of two 3.8-Ah cells connected in parallel, forming the 7.6-Ah block. As the capacity of blocks 1, 2, and 3 are the same, these blocks can be connected in series.

In the previous section, we learned that when a battery pack consists of multiple S and P configurations, P blocks are usually made first, and the blocks are connected in series later. Voltage of each cell or cell block needs to be monitored to avoid overcharging or overdischarging. When P blocks are made first, fewer voltage monitoring

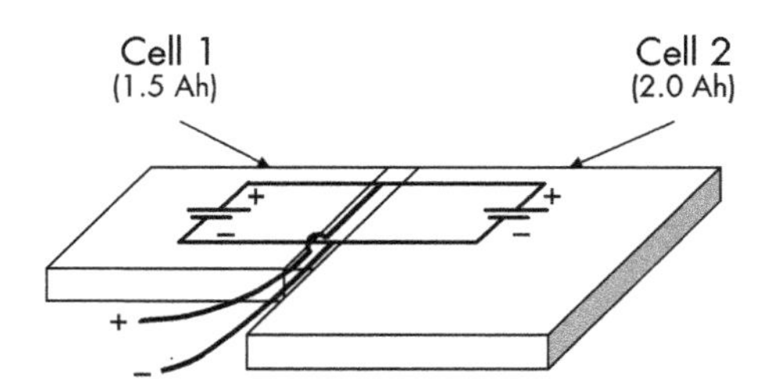


**Figure 6.20** An example of a nonrectangular 1S2P battery pack with two different-sized cells.

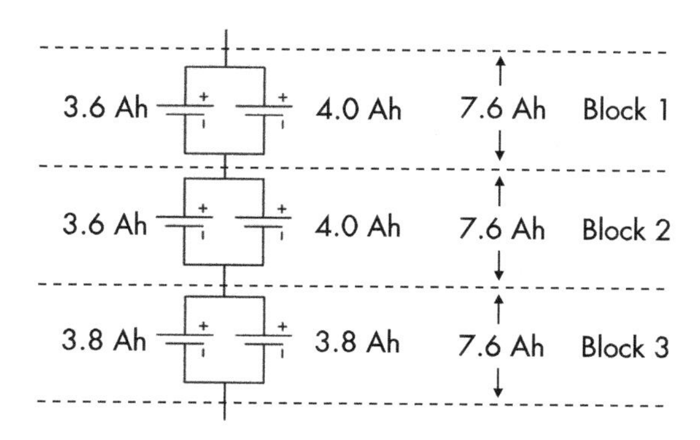


**Figure 6.21** An example of a nonrectangular 3S2P battery pack with three types of different-sized cells.

points are sufficient than the case when S is made first. For example, Figure 6.22 shows two kinds of 2S2P battery packs.

In Figure 6.22(a), 2P blocks are made first and the two blocks are connected in series. This requires only three connections for voltage measurement. However, when serial connection is made first as shown in Figure 6.22(b), four connections are required. Making P blocks first, as shown in Figure 6.22(a) is simpler and more affordable because of less voltage monitoring points. It is also safer because of less imbalanced cells.

In any cases of cell connection, careful inspection of current to and from each cell is needed to avoid overcurrent.

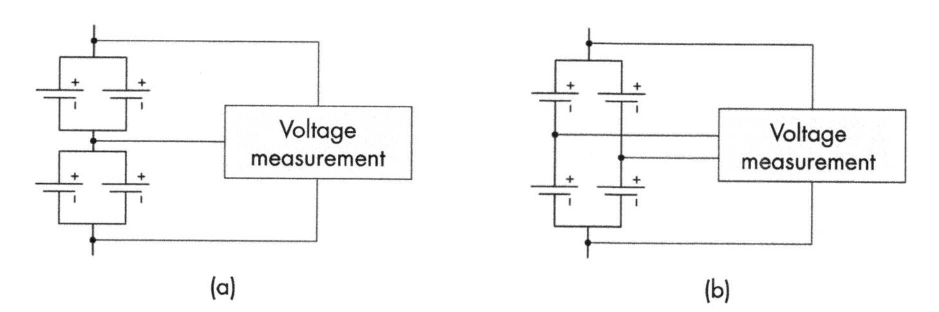


**Figure 6.22** Two kinds of 2S2P battery packs: (a) parallel connection first, and (b) serial connection first.

### 6.3.3 Shipping Regulations and Battery Certifications

For safe transport of Li-ion batteries, there are many shipping regulations, such as the International Civil Aviation Organization (ICAO) Technical Instructions, and the International Air Transportation Association (IATA) Dangerous Goods Regulations. At the time of writing this book, packing instructions for Li-ion batteries or Li-ion batteries contained/packed with equipment (e.g., a laptop PC with a battery pack) depend on the cell or pack Watt-hour rating. For example, when a cell exceeds 20 Wh or a battery (i.e., a battery pack with two or more cells) exceeds 100 Wh, a different packaging is required [9]. As this increases shipping cost, battery packs for portable devices (e.g., laptop PCs) are typically within 100 Wh. Also, when Li-ion rechargeable batteries are shipped by air without equipment, each battery SOC is limited to maximum 30% [9].

When Li-ion batteries are shipped, typically a UN38.3 test report, a drop-test report, and a safety data sheet (SDS) are required. UN38.3 is a standard written in Section 38.3 of the UN Manual of Tests and Criteria. It consists of eight tests, such as thermal test, shock test, and external short-circuit test.

In addition to shipping regulations, there are many other certifications and marking requirements. For example, IEC62133 is one of the major safety certifications for Li-ion battery cells and packs. Depending on the country, different certifications and recycling marks are required.

The regulations and/or certification requirements are updated often, and requirements are complicated. There are different rules for different chemistries, such as a lithium metal battery. A different country/airport/carrier may have different or additional rules. It is important to check the latest requirements and regulations for the battery and its transport route.

### 6.3.4 Authentication

Users may replace a battery pack in a system (e.g., smartphone) by themselves. If it is replaced with an inauthentic battery or a battery for a different system/model, that may cause a hazardous situation. For example, an inauthentic battery may have a different charging voltage requirement (e.g., 4.2V) from the authentic battery (e.g., 4.35V). If the system applies 4.35V charging to the inauthentic battery, that brings an overvoltage risk. If the authentic battery is capable of fast charging, the system may apply fast charging current to the inauthentic battery even if the inauthentic battery does not support fast charging. Also, the inauthentic battery not qualified by the system manufacturer may come with an improper/different design, poor quality, and other issues. For example, an authentic Li-ion battery pack for a battery-powered vacuum cleaner may not be high capacity but provides high power and long storage life. An inauthentic battery may show higher initial capacity that looks better to the users but may not provide as high power or as long a storage life.

To distinguish the authentic batteries from inauthentic batteries, or batteries for different models/systems, there are several identification (ID)/authentication methods. Some battery packs have a resistor and use it for ID. In this case, one of the pins in a connector of a battery pack is used to read the resistance. By the value of the resistance, ID or authentication is performed. While this method is easy to copy, it is a low-cost method.

Some battery packs have an IC that offers an authentication function, such as a fuel gauge IC. The IC communicates with the system and performs authentication (e.g., secure hash algorithm 1, which is SHA-1). This requires one or two pins in a connector of a battery pack.

### 6.3.5 Communication Protocol to Battery Pack

ICs in a battery pack communicate with the system and send/receive a lot of information. For example, a fuel gauge IC in a battery pack sends the information of the remaining battery capacity, full-charge capacity, SOC, battery temperature, pack voltage, cell voltage, current, and so forth. Some ICs also perform authentication. There are several communication protocols, such as system management bus (SMBus), and inter-integrated circuit (I2C), which is read as I-squared-C, and high-speed data queue (HDQ). SMBus is a derivative of I2C and is widely used in laptop PCs. Both SMBus and I2C require two pins in a connector of a battery pack for communication: one for data and the other for clock. HDQ requires only one pin in a battery connector, which saves space. However, the communication speed of two-wire protocols (e.g., 100 kHz) is more than ten times faster than that of HDQ. Therefore, two-wire protocols are preferred for the systems that frequently need to monitor the battery status. For example, a laptop PC battery typically uses two-wire protocol and carefully monitors battery voltage to avoid sudden system shutdown.

## 6.4 SUSTAINABILITY AND RECYCLING OF LI-ION BATTERIES

The production of Li-ion batteries has been drastically increasing, especially because of the demand for electric vehicles. It is important not only to meet the demand but also to enable the sustainable ecosystem, such as recycling. Sustainability is not the desire but should be the new normal. However, the ecosystem of Li-ion battery recycling is not sufficient yet. While the recycling process for a lead-acid battery is well established with up to 98% of the batteries able to be recycled, the recycling rate of Li-ion batteries is less than 5% [10, 11]. This section explains the sustainability of Li-ion batteries from three aspects: recycle, reuse, and reduce.

### 6.4.1 Recycle

The used Li-ion batteries are collected and sent to recycling facilities. Precious metals in lithium metal oxide cathodes such as lithium, cobalt, and nickel are recycled there. The major recycling methods are pyrometallurgical (pyro), hydrometallurgical (hydro), and combinations of these.

The pyrometallurgical method is basically to reduce the lithium metal oxide to mixed metals by heating it under vacuum or inert atmosphere. In the hydrometallurgical method, metals in batteries are dissolved into aqueous acids and precipitated selectively [12].

While the recycling methods are established, the low-recycling rate needs to be improved. One of the solutions is to standardize the spec and dimensions of Li-ion batteries (e.g., standardized EV batteries). This enables efficient transport and recycling of used batteries, resulting in cost reduction, more profitable recycling business, and improvement of recycling rate.

### 6.4.2 Reuse

Reuse of used Li-ion batteries is another way to enhance sustainability. When a Li-ion battery shows replacement notification through a system, the full-charge capacity of the battery may still be 70% to 80%

of the initial full-charge capacity. It may be sufficient for some other applications, such as stationary or backup batteries. To promote/enable the reuse, standardization of the spec and dimensions is desired across products.

### 6.4.3 Reduce

Reduction of battery disposal also contributes to sustainability. One of the solutions is to extend the battery longevity by battery control algorithms. Li-ion batteries degrade faster when batteries are charged to high SOC, such as full-charging. Fast charging or fast discharging also accelerates the degradation. By avoiding these degradation situations by algorithms, battery longevity can be extended [13–15]. For example, unnecessary full charging or fast charging can be avoided by using system software where the user can manually lower the maximum charging level and/or charging speed. With the aid of machine learning algorithms, the setting change can be automatic. The details are explained in Chapters 10 and 12.

## 6.5 SUMMARY

In this chapter, we learned the following:

- Li-ion battery cell and pack structure/configuration.
- Manufacturing process.
- Connection of different-sized cells.
- Methods to enhance sustainability.

This covers the entire flow of Li-ion battery products. Often, one battery job focuses on one specific area. However, by understanding the entire flow, better achievement of the job is possible, for example, battery cell/pack design where the materials can be easily recycled. Such entire optimization is desired for the next-generation batteries.

## 6.6 PROBLEMS

### Problem 6.1

There is a battery engineer in a smartphone design team. According to a system designer, the battery space in a new smartphone design is $4.0 \times 40.5 \times 80.0$ mm. To maximize battery life, the battery engineer asked a battery pack supplier to make a pouch battery pack with $4.0 \times 40.5 \times 80.0$ mm. What would happen?

Answer 6.1

A Li-ion battery swells after cycles (e.g., 8% in thickness). Clearance for the swelling needs to be allocated for safety. Therefore, the initial battery needs to be thinner than 4.0 mm.

Problem 6.2

There are six 1-Ah cells. Each cell is nominal 3.8V and the impedance is 30 mohm. When the six cells are connected and a 3S2P battery pack is formed, what is the voltage, capacity, and impedance of the battery pack?

Answer 6.2

Nominal pack voltage is $3.8V \times 3 = 11.4V$. Pack capacity is 2 Ah because each 2P block is $1 \text{ Ah} \times 2 = 2$ Ah. Pack impedance is $(30 \text{ mohm}/2) \times 3 = 45$ mohm.

Problem 6.3

There is a 2S3P battery pack where each cell has 19 Wh. What is the concern with using the battery pack for a laptop PC?

Answer 6.3

The 2S3P battery pack has 6 cells. When one cell has 19 Wh, the battery pack has $19 \text{ Wh} \times 6 = 114$ Wh. This exceeds a 100-Wh limit in the IATA dangerous goods regulations and special packaging is required, which may cost more.

Problem 6.4

There are two cells. One is 2 Ah and the other is 1.9 Ah. Charge and discharge cutoff voltages of the two cells are the same.

1. When these are connected in series, how much capacity can be provided at maximum in a safe manner?

2. When these are connected in parallel, how much capacity can be provided at maximum in a safe manner?

Answer 6.4

1. The answer is 1.9 Ah. As the cells are connected in series, the same Ah needs to be provided from each cell. Exceeding 1.9 Ah causes overdischarging in the 1.9-Ah cell.

2. The answer is 3.9 Ah. Different-capacity cells can be connected in parallel when charge and discharge cutoff voltages are the same.

## References

[1] Liu, Y., et al., "Current and Future Lithium-ion Battery Manufacturing," *iScience*, Vol. 24, No. 4, 2021, p. 102332.

[2] Smekens, J., et al., "Influence of Electrode Density on the Performance of Li-Ion Batteries: Experimental and Simulation Results," *Energies*, Vol. 9, 2016, p. 104.

[3] Sun, Y., "Promising All-Solid-State Batteries for Future Electric Vehicles," *ACS Energy Letters*, Vol. 5, No. 10, 2020, pp. 3221–3223.

[4] Bates, J. B., et al., "Thin-Film Rechargeable Lithium Batteries," *Journal of Power Sources*, Vol. 54, No. 1, 1995, pp. 58–62.

[5] Juarez-Robles, D., et al., "Overdischarge and Aging Analytics of Li-Ion Cells," *Journal of the Electrochemical Society*, Vol. 167, No. 9, 2020, p. 090558.

[6] Ma, S., et al., "Temperature Effect and Thermal Impact in Lithium-ion Batteries: A Review," *Progress in Natural Science: Materials International*, Vol. 28, No. 6, 2018, pp. 653–666.

[7] Mao, Z., et al., "Calendar Aging and Gas Generation in Commercial Graphite/NMC-LMO Lithium-Ion Pouch Cell," *Journal of the Electrochemical Society*, Vol. 164, No. 14, 2017, pp. A3469–A3483.

[8] Zhang, S., "Insight into the Gassing Problem of Li-ion Battery," *Frontiers in Energy Research*, Vol. 2, 2014.

[9] International Air Transport Association (IATA), "Lithium Batteries," https://www.iata.org/en/programs/cargo/dgr/lithium-batteries/.

[10] Ballantyne, A., et al., "Lead Acid Battery Recycling for the Twenty-First Century," *Royal Society Open Science*, Vol. 5, Issue 5, 2018, p. 171368.

[11] "Recycle Spent Batteries," *Nature Energy*, Vol. 4, No. 4, 2019, p. 253.

[12] Baum, Z., et al., "Lithium-Ion Battery Recycling Overview of Techniques and Trends," *ACS Energy Letters*, Vol. 7, No. 2, 2022, pp. 712–719.

[13] Matsumura, N., "Battery Cycle Life Extension by Charging Algorithm to Reduce IOT Cost of Ownership," *The 34th International Battery Seminar & Exhibit*, Florida, 2017.

[14] Matsumura, N., et al., "A Method to Enhance System Peak Power While Mitigating Battery Degradation," *The 37th International Battery Seminar & Exhibit*, Florida, 2020.

[15] Matsumura, N., et al., "Context-based Battery Charging Algorithm, an Application of Machine-Learning/Deep-Learning to Battery Charging for Longevity Extension," *The 39th International Battery Seminar & Exhibit*, Florida, 2022.

# 7

# BATTERY FUEL GAUGING METHODS

## 7.1 INTRODUCTION

The systems that use batteries show a charge level indicator. For example, smartphones and laptop PCs typically display a battery icon that graphically shows the battery SOC. Some systems may also display the remaining battery life in hours and minutes. Electric vehicles display the remaining driving range. How are these estimated? The measurement and estimation of the remaining battery capacity are called fuel gauging (or sometimes gas gauging). There are several fuel gauging methods ranging from a simple model to a machine-learning-based algorithm. The fuel gauging method is performed by a fuel gauge IC or an IC that has a fuel gauging algorithm in its software/firmware. This chapter explains the fuel gauging methods for Li-ion batteries.

## 7.2 VOLTAGE MEASUREMENT

This section explains the fuel gauging method with voltage measurement.

### 7.2.1 Theory

Battery OCV and SOC are correlated. This method measures battery voltage and identifies the corresponding SOC. The following explains the reason why OCV corresponds to SOC.

When a Li-ion battery consists of $LiCoO_2$ cathode and graphite anode as an example, the cathode formula changes between $Li_{1-x}CoO_2$ and $LiCoO_2$, depending on SOC. The anode formula also changes by SOC between $LiC_6$ and $C_6$ through several intermediate lithium-carbon stages [1].

As explained in Chapter 2, each substance has its own Gibbs free energy that leads to its own voltage through the Nernst equation. The difference between the cathode and the anode voltages is the battery voltage. Figure 7.1 shows how the cathode, the anode, and the battery voltage changes over SOC.

As cathode ($LiCoO_2$) and anode (graphite) voltages correspond to SOC, the battery voltage, which is the difference between cathode and anode voltages ($LiCoO_2$ versus graphite), corresponds to SOC as well. This means that SOC can be determined by a proper measurement of battery OCV.

Figure 7.2 shows an example of SOC and battery OCV relationship.

To estimate the SOC, first read the OCV of the battery, then read the corresponding SOC as shown in Figure 7.2. In a real system, a fuel

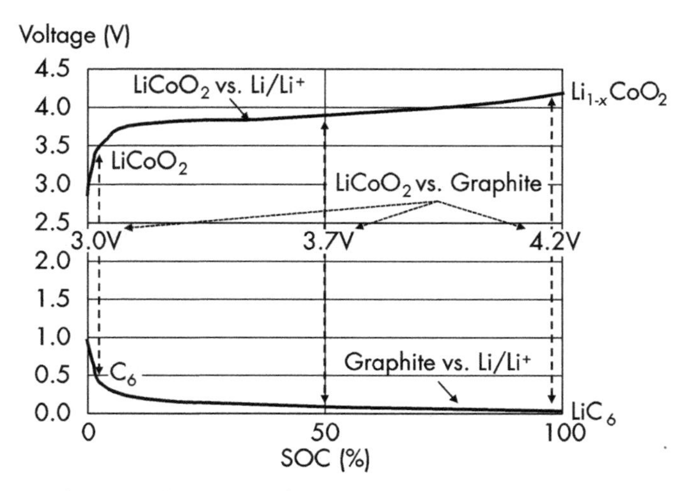


**Figure 7.1** Schematic illustration of $LiCoO_2$ and graphite voltages relative to $Li/Li^+$ over SOC.

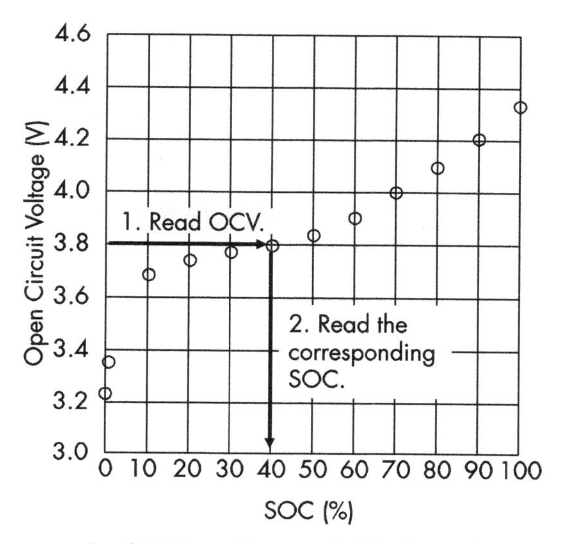


**Figure 7.2** An example of SOC and battery OCV relationship.

gauge IC or fuel gauging algorithm has a formula or a lookup table that shows the relationship between OCV and SOC. It first reads the battery voltage, then estimates the battery SOC by the formula or the table.

***Exercise***

1. When a battery OCV is 4.2V, what is the battery SOC? Use Figure 7.2.
2. When a battery OCV is 3.8V, what is the battery SOC? Use Figure 7.2.

***Answer***

1. 4.2V corresponds to ~90% SOC.
2. 3.8V corresponds to ~40% SOC.

* * *

As we noticed, ~40% is in a relatively flat region of the curve. A small error in voltage measurement causes a large error in SOC estimation.

The OCV-SOC relationship depends on the cathode and anode choices because different cathode/anode has different Gibbs free energy. It also depends on temperature because the Gibbs free energy is defined as a function of temperature.

Figure 7.3 compares the OCV-SOC relationship between a new battery and a 500-cycled battery.

When a battery is properly designed and manufactured with good quality, such as the battery in Figure 7.3, the OCV-SOC relationship does not largely change between initial and after cycles as shown in the figure. However, if one of cathode or anode is degraded faster than the other, voltage curves of cathode and anode in Figure 7.1 are aligned differently, which results in the change of a OCV-SOC relationship.

Note that while the OCV-SOC relationship of a good quality battery does not largely change over cycles, the absolute capacity represented by 100% SOC does decrease. More on this is explained in Section 7.3.

### 7.2.2 Advantages and Disadvantages

To perform this method, voltage measurement and simple conversion software/firmware are required. While this method is low cost, there

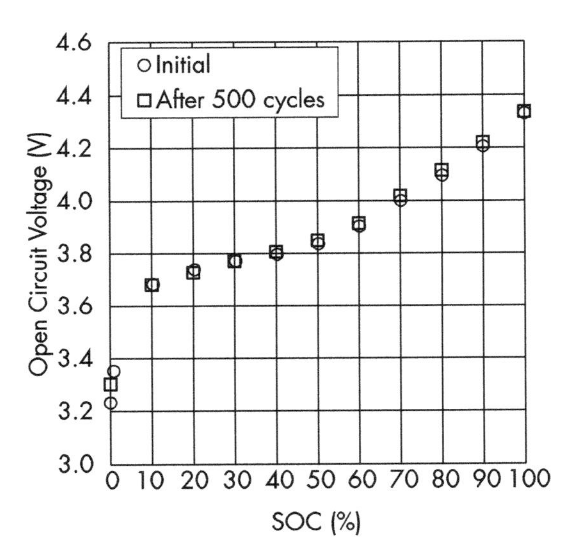


**Figure 7.3** Comparison of OCV-SOC relationship between before and after 500 cycles.

are several challenges. For example, this method is less accurate than the other fuel gauging methods that are explained later. In a system, the battery voltage is influenced by discharging or charging current. The measured voltage may include IR drop (or IR jump) and may not be OCV. To measure OCV, it takes a long time to rest the battery under open circuit and reach the true OCV, which is in the equilibrium state. Measuring the battery voltage before reaching OCV includes the influence by IR and causes an error in SOC estimation. Figure 7.4 is an example of an SOC estimation error due to IR drop.

In this figure, there are two voltage-SOC curves. One is at OCV and the other is during 0.5C discharge. Battery voltage during 0.5C discharge is lower than that at OCV because of IR drop. When the measured voltage is 3.75V and the voltage is OCV, 3.75V reading corresponds to ~25% SOC. However, when the battery is delivering 0.5C, the same voltage reading corresponds to ~55% SOC, which is a large error. The error due to IR drop is affected not only by current but also by temperature as impedance changes with temperature. The temperature influence can be especially pronounced at low temperature.

In this example, the voltage is read during discharge. However, note that even if the current is zero, the voltage may not represent the

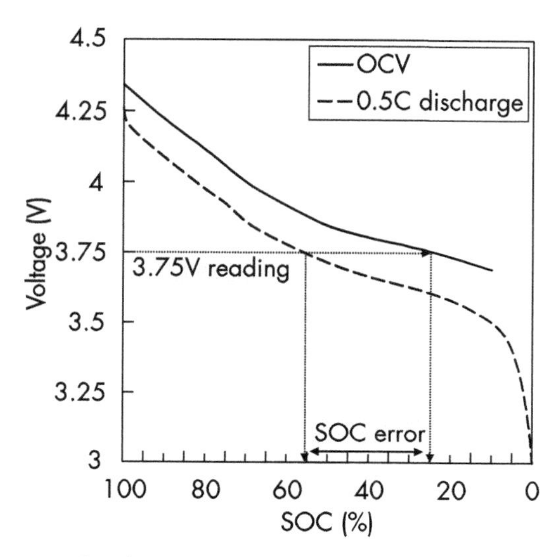


**Figure 7.4** An example of an SOC difference at the same voltage.

true OCV. It takes time for the battery to relax to equilibrium after any change in current, charge, or discharge.

Also, a small reading error in the battery voltage may lead to a several-percent error in SOC, especially in the relatively flat region of the OCV-SOC curve.

Furthermore, this method does not estimate capacity, while it estimates SOC. The full-charge capacity decreases over cycles and time due to degradation. SOC shows only the state of charge as percentage relative to the decreased full-charge capacity. It does not show the remaining battery capacity, the indicator of the remaining battery life, because the decreased full-charge capacity is unknown. For example, if the full-charge capacity of the battery is 2 Ah, the remaining capacity is 1 Ah (i.e., 50% SOC) and the system runs at 1A, 1 Ah/1A = 1 hour battery life is expected as the remaining battery life. However, if the fuel gauge does not know the full-charge capacity and only estimates the SOC as 50%, the remaining battery life cannot be estimated.

Still, this method is simple and low cost. Some methods using voltage measurement may read the battery voltage early before reaching OCV and estimate OCV and the corresponding SOC tolerating the errors, which saves time. Therefore, it is suitable for systems that are low power and do not require accurate fuel gauging. In this method, the battery charge level is typically displayed as a multilevel bar indicator, such as three levels that indicate high (3-level), medium (2-level), and low (1-level).

## 7.3 COULOMB COUNTING

### 7.3.1 Theory

The coulomb counting method counts the coulombs passed through the battery. As 1 coulomb per second equals 1 ampere, counting coulombs over time shows cumulative ampere-hour. Figure 7.5 is an example of coulomb counting implementation.

In this figure, coulomb counting consists of a current sense resistor and a fuel gauge IC. When the resistance of the resistor is known as $R$, the fuel gauge IC measures the voltage drop ($dV$) at the resistor and calculates current by $dV/R$. The continuous measurement of current to and from the battery over time shows how much Ah is charged to or discharged from the battery.

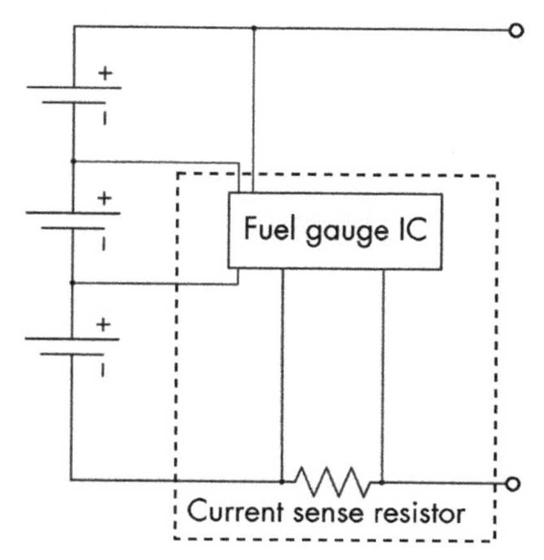


**Figure 7.5** An example of a coulomb counting implementation.

***Exercise***

1. When 1A is passed to the battery for 1h, how much Ah is charged?
2. When 1A is drawn from a fully charged 3-Ah cell for 1h, how much Ah is remaining?
3. When 0.5A is passed to an empty cell for 1h and then 0.2A is drawn for 0.5h, how much is the remaining battery capacity?

***Answer***

1. $1A \times 1h = 1$ Ah
2. 3 Ah − 1 Ah = 2 Ah
3. $0.5A \times 1h - 0.2A \times 0.5h = 0.4$ Ah

### 7.3.2 Advantages and Disadvantages

This method shows capacity (Ah), which leads to the estimation of the remaining battery life. Also, when the full-charge capacity is known as $Q_{full}$, measured remaining capacity ($Q$) shows the SOC of the battery by $(Q/Q_{full}) \times 100$.

However, counting errors happen when current is below the detection limit, such as a too short pulse current or too low current. It cannot detect the capacity change either when the capacity is consumed without the current going through the sense resistor, such as self-discharge in a cell. Error can also come from noise on the measurement of the sense resistor. Such errors accumulate over time.

Batteries degrade over time and the full-charge capacity decreases. If the battery is charged from empty to full at a detectable current, the full-charge capacity can be recalibrated. However, such recalibration requires full discharging first, followed by full charging, which takes time and is not convenient to the users.

Also, if the IR drop, which is a function of current and temperature, is not considered, the error increases. For example, when a system runs at 0.5A and some capacity (e.g., 0.5 Ah) remains in the battery of the system, the system may be expected to run for another 1h. However, if the system draws higher current and/or the battery temperature is low, the battery voltage hits the system shutdown voltage earlier due to larger IR drop. In this case, battery life estimation is not accurate because part of the remaining 0.5 Ah is unused. Figure 7.6 is an example of usable capacity varying with temperature at 0.2C.

In this figure, discharging at 25°C provides the most capacity. When the temperature decreases, the discharged capacity decreases because IR drop is larger at lower temperature and the system shutdown voltage is hit earlier. This illustrates how important it is to consider IR drop for fuel gauging.

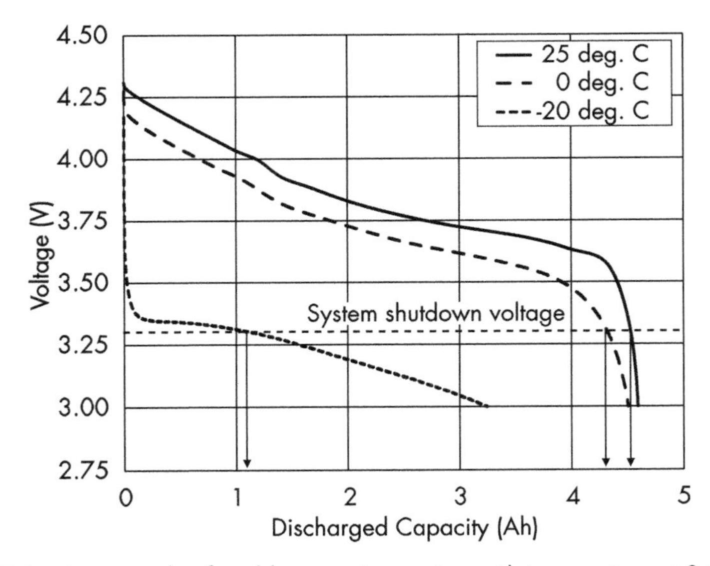


**Figure 7.6** An example of usable capacity varying with temperature at 0.2C.

## 7.4 VOLTAGE MEASUREMENT AND COULOMB COUNTING

### 7.4.1 Theory

This method combines the two techniques explained earlier: voltage measurement and coulomb counting. Figure 7.7 shows the procedure for this method.

*Step 1:* When the system is idle and battery voltage is OCV at point 1, the battery voltage is measured and the corresponding SOC, which is $X$ %, is recorded.

*Step 2:* The battery is charged to point 2. During charge, the coulomb counter counts the coulombs (i.e., capacity, $Q_{chg}$) passed between points 1 and 2.

*Step 3:* When the system is idle for a while and battery voltage is OCV at point 2, the battery voltage is measured and the corresponding SOC, which is $Y$ %, is recorded.

*Step 4:* Full-charge capacity ($Q_{FCC}$) is calculated as:

$$Q_{FCC} = [Q_{chg}/(Y - X)] \times 100 \tag{7.1}$$

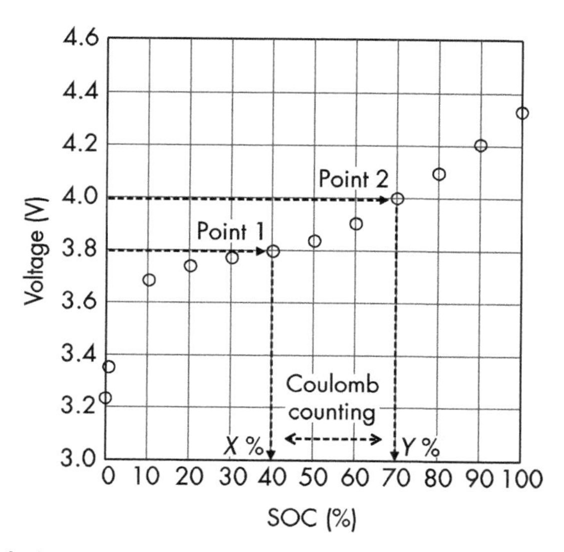


**Figure 7.7** A fuel gauging method of voltage measurement and coulomb counting.

***Exercise***

When $Q_{chg}$ is 1.0 Ah in Figure 7.7, what is the full-charge capacity?

***Answer***

From the figure, $X$ is 40% and $Y$ is 70%. $Q_{FCC}$ = [1.0 Ah/(70% – 40%)] × 100 ≈ 3.3 Ah.

* * *

While Figure 7.7 shows the method during charge, the same method can be applied during discharge as well.

### 7.4.2 Advantages and Disadvantages

This method updates the full-charge capacity faster than coulomb counting without full discharging and full charging for recalibration. This is especially useful for systems where the battery is usually not fully discharged, such as electric vehicles and backup batteries.

While this method resolves one of the disadvantages of coulomb counting, it still relies on being able to make an accurate OCV measurement at the beginning and end of the coulomb counting period. Impedance impact needs to be considered to estimate the usable battery capacity and the remaining battery life.

## 7.5 IMPEDANCE CONSIDERATION

### 7.5.1 Theory

For OCV and battery-life estimation, some models consider battery impedance [2]. Such models own the battery impedance database as a function of SOC, current level, and temperature. During charge, the measured voltage $V$ is not OCV due to the influence of the IR jump. If impedance is known, OCV can be calculated as $V - I \times R$ where $I$ is charging current, and $R$ is battery impedance. This enables faster OCV estimation, resulting in faster estimation of the full-charge capacity.

When the battery is degraded, battery impedance increases. Therefore, it is required to monitor voltage change during discharge or charge, recalculate impedance, and periodically update the impedance database. For example, impedance is recalculated by $|V_{measured} - V_{OCV}|/I$, where $V_{measured}$ is the voltage measured after a certain time under current $I$ and $V_{OCV}$, is OCV. For the details of impedance, refer to Chapters 3 and 4.

Impedance consideration can also make capacity and battery-life estimation more accurate. Figure 7.8 is an example.

During discharge, battery voltage drops due to IR. In Figure 7.8, when current increases, the voltage under low current ($V_1$) drops to $V_2$, which is the voltage under high current. $V_2$ can be expressed as:

$$V_2 = V_1 - \mathrm{d}I \times R \tag{7.2}$$

where d$I$ is the increase in discharging current from $V_1$ to $V_2$, and $R$ is battery impedance.

Under low current, the fuel gauging model predicts that the discharge curve hits the system shutdown voltage at point b, and the capacity at point b′ is usable. When high current is detected, the fuel gauging model considers IR drop, updates the discharge curve, predicts that the new curve hits the system shutdown voltage at point a, and estimates that the capacity at point a′ is usable.

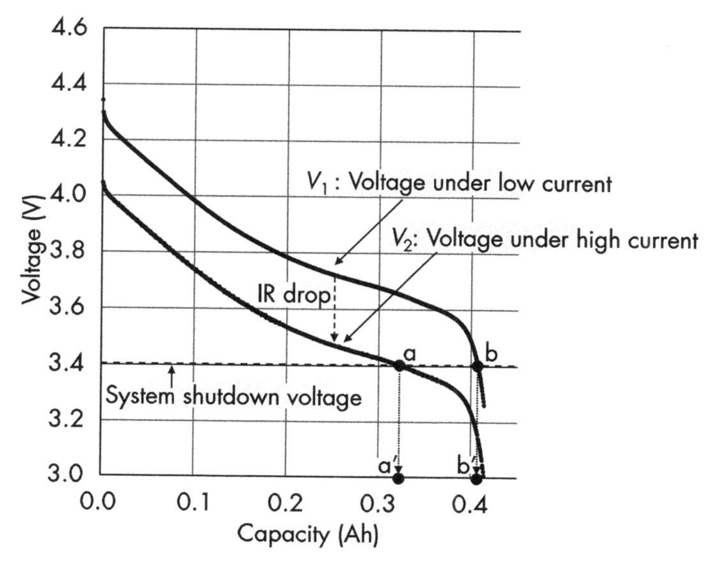


**Figure 7.8** An example of usable-capacity estimation considering impedance.

### 7.5.2 Advantages and Disadvantages

By considering impedance, faster OCV estimation is possible, which leads to faster update of the full-charge capacity. Also, estimation of usable battery capacity becomes more accurate, such as within a 1% error on ideal conditions where current is low/stable. However, as explained in Chapter 3, battery impedance includes not only resistance but also capacitive reactance, which is influenced by the charging and discharging history. This means that IR drop depends on the duration of the current as well as magnitude of current, temperature, SOC, and degradation status. The actual accuracy is largely influenced by the usage. For example, when a system is under the sporadic burst-power mode that intermittently draws high current from the battery for a short period, battery voltage in Figure 7.8 changes between $V_2$ and $V_1$. In this case, $R$ in (7.2) is not a constant value, but a function of current duration. Longer burst mode causes larger IR drop. If a fuel gauge IC considers the worst-case IR drop (e.g., point a in Figure 7.8) to estimate the usable capacity, but the IR drop is not as large, the system may run longer than expected. This is because the end of the capacity is estimated as point a, but it is actually in between a and b. If the impedance estimation is optimistic but the real IR drop is larger, the system shuts down suddenly. To avoid this, some systems, such as smartphones and electric vehicles, regulate power consumption by, for example, dimming the smartphone screen, or limiting the speed of the electric vehicle when SOC is low. This reduces IR drop and the usable capacity can be extended.

## 7.6 ADVANCED FUEL GAUGING EXAMPLES

OCV and SOC are correlated. For fuel gauging models, fast and accurate OCV prediction is desired for fast SOC and battery-life estimation. When battery charging or discharging stops, the battery voltage starts going back to OCV, but it takes time because of the capacitive reactance in the battery. It is important to predict OCV immediately after charging or discharging stops. This section explains some examples of the OCV/SOC prediction models and the importance of impedance estimation.

### 7.6.1 OCV Prediction with an Equivalent Circuit Model

Chapter 3 explained that battery impedance can be explained with an equivalent circuit model such as in Figure 7.9.

This figure consists of one resistor and two parallel RC circuits, $V_0$ is OCV. When battery charging or discharging stops, IR drop at $R_0$ becomes zero immediately. IR drops at two parallel RC circuits decrease towards zero but this takes a while. In this model, the battery voltage recovery to OCV after discharging can be expressed as:

$$V_{batt} = V_0 - V_{ini,1} \exp[-t/(R_1C_1)] - V_{ini,2} \exp[-t/(R_2C_2)] \quad (7.3)$$

where $V_{batt}$ is the observed battery voltage at time $t$, $V_0$ is OCV, $V_{ini,1}$ is the initial voltage of the parallel $R_1C_1$ circuit, $V_{ini,2}$ is the initial voltage of the parallel $R_2C_2$ circuit, $R_1$ and $R_2$ are resistance, and $C_1$ and $C_2$ are capacitance. It is difficult to know the exact values of $V_{ini,1}$, $V_{ini,2}$, $R_1$, $R_2$, $C_1$, and $C_2$. However, monitoring battery voltage $V_{batt}$ for a while during the voltage recovery, and fitting the model to the data gives information to predict OCV. For example, (7.3) can be rewritten as:

$$V_{batt} = V_0 - a[\exp(-t/b)] - c[\exp(-t/d)] \quad (7.4)$$

where $a$, $b$, $c$, and $d$ are the constant parameters.

What would happen if this equation were fitted to the data? Figure 7.10 is an example of battery voltage recovery after discharging.

This figure shows both battery voltage and current. The battery was discharged to point 1 and 1′. When discharging stops, the battery

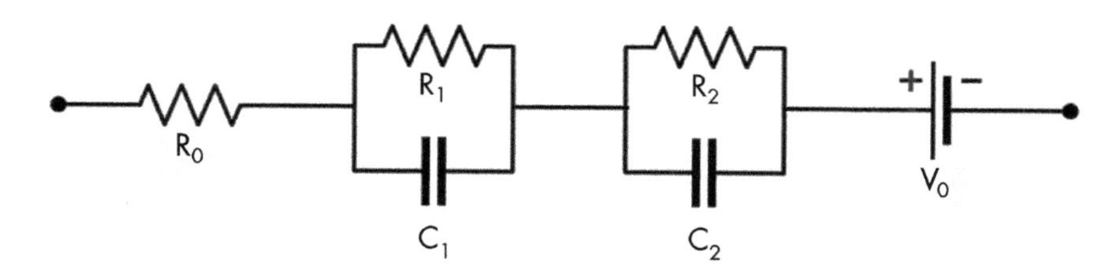


**Figure 7.9** An example of a battery equivalent circuit model with one resistor and two parallel RC circuits.

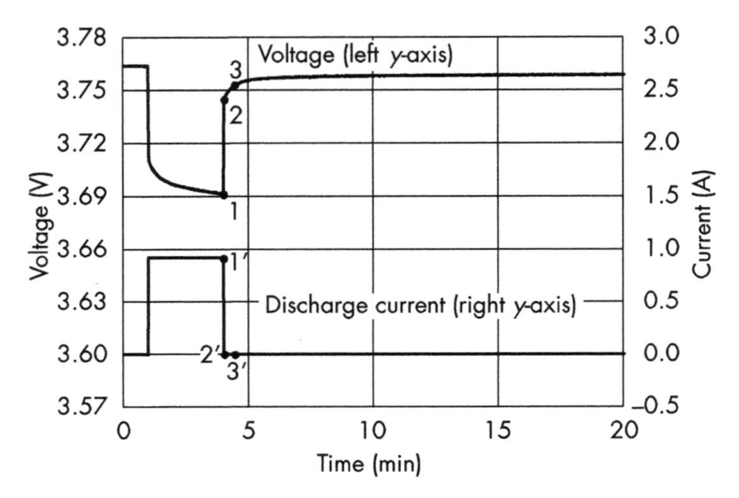


**Figure 7.10** An example of battery voltage recovery after discharging.

voltage comes back to point 2 immediately and keeps increasing towards OCV. In this case, the measured OCV is 3.759V. When (7.4) is fitted to the 30-second data between point 2 and 3 for example, the following equation is obtained:

$$V_{batt} = 3.756 - 5.731 \times 10^{-4} \times [\exp(-t / 0.05406)] \\ -1.055 \times 10^{-2} \times [\exp(-t / 0.3696)] \tag{7.5}$$

OCV is the voltage when sufficient rest time is given. Therefore, when $t$ goes to infinity, the second and third terms on the right side of (7.5) go to zero and $V_{batt}$ goes to OCV, which equals 3.756V. This is close to the measured OCV (3.759V).

This is one example of OCV predictions with equivalent circuit models. While the example is based on an equivalent circuit model, it performs fitting and OCV prediction based only on the measured voltages. It does not require resistance and capacitance measurements in advance or during operation. The prediction can be better when the model is further optimized with more parameters and/or more data is available for fitting. However, the fitting for complex equivalent circuit models with many parameters may be computationally expensive in some systems.

On a side note, this section explained an example of OCV estimation after discharging. In case of OCV estimation after charging with the equivalent circuit model in Figure 7.9, the voltage recovery to OCV after charging stops is expressed as follows:

$$V_{batt} = V_0 + V_{ini,1} \exp[-t/(R_1C_1)] + V_{ini,2} \exp[-t/(R_2C_2)] \quad (7.6)$$

For parameter fitting, the following equation is used:

$$V_{\mathrm{batt}} = V_0 + a[\exp(-t/b)] + c[\exp(-t/d)] \quad (7.7)$$

The difference between (7.4) and (7.7) is the signs of the second and the third terms on the right side of the equation. This is because RC parallel circuits in Figure 7.9 are reversely charged and the measured voltage is higher than OCV due to IR jump.

### 7.6.2 SOC Prediction with Machine Learning

SOC can be predicted by voltage and impedance, which is the function of battery SOC, current, temperature, and so forth. While the function is complex, there are approaches to use machine-learning models or deep-learning models that are part of the machine-learning models [3]. For example, with a neural-network model, which is one of the machine-learning models, measured data of instantaneous voltage and temperature with moving average current and moving average voltage are connected to SOC through many conversion layers [4]. These layers are called a neural network. All conversion parameters in the conversion layers are fine-tuned with training data. Once it's done, the model can predict SOC under different voltage, temperature, moving average current, and moving average voltage.

Chapter 12 explains the overview of machine-learning/deep-learning and its application to battery charging.

### 7.6.3 Power Optimization Considering Battery Impedance

When battery SOC is low, some systems may reduce discharging current and power to mitigate IR drop and delay hitting the system shutdown voltage. In such a situation, the following equation needs to be met to avoid sudden system shutdown:

$$V_{batt} = V_{OCV} - I \times R \geq V_{minsys} \tag{7.8}$$

where $V_{batt}$ is measured battery voltage, $V_{OCV}$ is battery OCV, $I$ is discharging current, $R$ is battery impedance, and $V_{minsys}$ is the system shutdown voltage.

When impedance $R$ is overestimated and current $I$ is thus largely limited, system response may become unnecessarily slow and/or system functions are unnecessarily limited, resulting in worse user experience. However, when impedance $R$ is accurately and promptly estimated, current $I$ does not need to be reduced largely [5]. That enhances the user experience when the battery is at low SOC.

Battery impedance is complex. However, knowing it opens the door for better fuel gauging, longer battery life, and enhanced performance.

## 7.7 STATE OF HEALTH

When full-charge capacity of the battery is known, battery state-of-health (SOH) can be calculated as:

$$SOH = (Q_{FCC}/Q_{FCC,\ ini}) \times 100 \tag{7.9}$$

where $Q_{FCC}$ is the present full-charge capacity and $Q_{FCC,ini}$ is the initial full-charge capacity when the battery was new. Figure 7.11 is an example of battery cycle life and SOH.

When the battery is new at cycle 0, the initial capacity is 3.36 Ah. After 500 cycles, the full-charge capacity decreases to 3.07 Ah. In this case, SOH is 3.07 Ah/3.36 Ah × 100 = 91.4%.

SOH indicates when the users are recommended to replace the battery, such as 80% SOH. For consumer electronics devices such as smartphones and laptop PCs, 80% or higher SOH is typically expected after 500 to 1,000 cycles at standard charging and discharging conditions, such as 0.5C at room temperature. Note that in a real system, decrease in the full-charge capacity is influenced not only by the number of cycles but also by many other factors, such as magnitude of charging/discharging current, battery temperature, and SOC during storage or operation. Accurate fuel gauging leads to accurate SOH estimation.

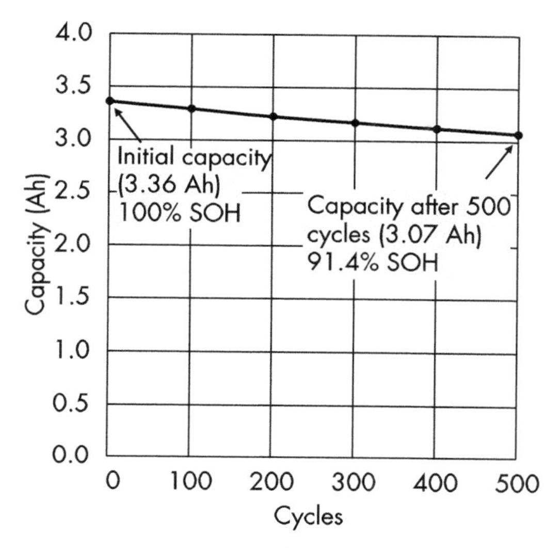


**Figure 7.11** An example of battery cycle life and SOH.

## 7.8 SYSTEM-SIDE FUEL GAUGE VERSUS PACK-SIDE FUEL GAUGE

The fuel gauging method can reside in a fuel gauge IC in a battery pack or an IC on a system motherboard. While portable devices such as smartphones and laptop PCs typically use pack-side fuel gauge ICs, there are some systems that use system-side fuel gauges, for example, by implementing a fuel gauging algorithm in the microcontroller on the motherboard. Figure 7.12 shows examples of a system-side fuel gauge and a pack-side fuel gauge.

If system-side fuel gauge is enabled by implementing a fuel gauging algorithm in an existing IC on a system, that saves cost because a fuel gauge IC in a battery pack is not needed. Also, when a high-performance IC is used for fuel gauging, more complex algorithms, such as deep-learning-based algorithms, can be used. However, such ICs on the system side may consume more power than the dedicated fuel gauge ICs in a battery pack. The fuel gauging is always running unless the system is completely off. In other words, the fuel gauging is always performed when the system is under operation or is sleeping. If the ICs consume more power, battery life is negatively impacted.

When the system-side fuel gauge IC monitors the battery voltage, it needs to consider additional impedance, including contact resistances in a connector between a system and a battery pack. If the

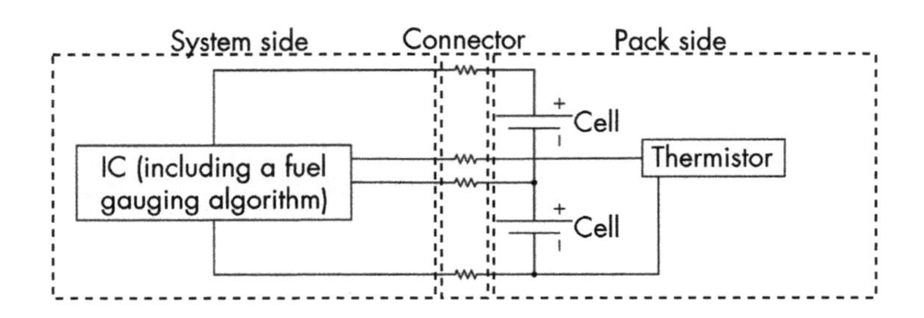


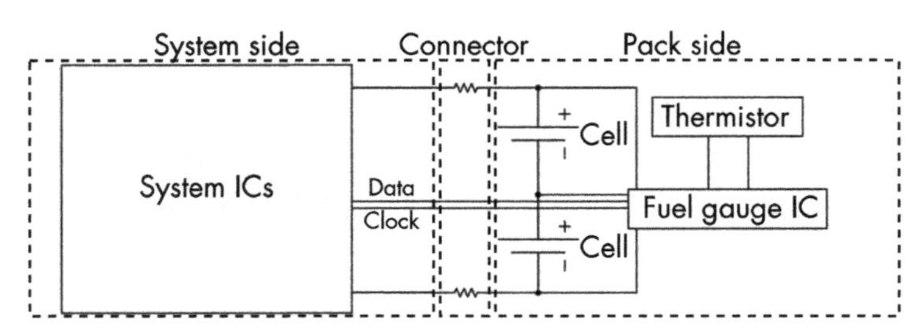


**Figure 7.12** An example of system-side fuel gauge (top) vs pack-side fuel gauge (bottom).

impedance varies by systems or changes after battery replacement, for example, due to variance in contact resistance in a connector, the fuel gauging is less accurate.

Also, system-side fuel gauge may additionally need to integrate safety protections that the pack-side fuel gauge ICs usually offer, such as cell balancing and the protections from over-/under-voltage, overcurrent, over-/undertemperature, and short circuit. If impedance between a system and a battery varies by systems, that may risk accurate protection. When a battery pack consists of multiple S cells or cell blocks such as 3S, voltage of each cell or cell block needs to be monitored. To perform this from the system side, additional pins in the connector may be needed, which takes additional space and increases cost for the connector.

## 7.9 SUMMARY

In this chapter, we learned about several fuel gauging algorithms, from the simple, but less accurate, to the complex but more accurate. Advanced fuel gauging algorithms characterize batteries under various SOC, current and temperature conditions over time, depending on battery designs from multiple battery suppliers for a given system. The fuel gauging algorithms consist of telemetry, such as voltage measurement and coulomb counting, and battery knowledge such as electrochemistry and equivalent circuit models. Recently, new algorithms have been researched with machine learning such as SOC estimation. Also, new battery usage, such as batteries for electric vehicles, may require new algorithms. For example, the driving range of electric vehicles may need to be estimated by considering the temperature on the drive since battery temperature affects usable capacity. As batteries are used in more applications where machine learning may be available, more opportunities are coming.

## 7.10 PROBLEMS

### Problem 7.1

Refer to Figure 7.7. There is a fuel gauge IC that has both voltage measurement and coulomb counting. At a point, the IC measured a cell and read 3.8V as OCV. After 2.0-Ah charging, the IC measured the cell again and read 4.2V as OCV. What is the full-charge capacity of the cell?

### Answer 7.1

$Q_{chg}$ is 2 Ah. From the figure, 3.8V and 4.2V correspond to 40% and 90%, respectively. $Q_{FCC}$ = [2.0 Ah / (90% – 40%)] × 100 = 4.0 Ah.

### Problem 7.2

You were watching a movie on a smartphone where the battery indicator showed 10%. When you stopped the movie and switched the application to a 3-D video game, the smartphone shut down suddenly. What is the possible reason?

### Answer 7.2

The fuel gauge originally showed 10% SOC by predicting the usable capacity based on the IR drop while watching a movie. However, when the 3-D video game started, it drew higher current from the battery, which caused a larger IR drop. This resulted in hitting the system shutdown voltage earlier.

Problem 7.3

You own an electric vehicle. One day in winter, you decided to drive to the ski resort, which is 100 miles away. You confirmed that the remaining driving range of the car is 150 miles, which should be sufficient to arrive there. However, the battery of the car ran out on the way there. What were the possible reasons?

Answer 7.3

(Example 1) The fuel gauge of electric vehicles may be based on a normal driving speed in a city. However, while driving on a highway at higher speed and/or driving uphill, higher current was drawn from the battery, which caused larger IR drop, resulting in shorter driving range. If an electric car heater was on, it drew additional current and caused even shorter driving range.

(Example 2) Battery impedance increases at low temperature. When you departed, the battery temperature might have been normal. However, it decreased when you neared the ski resort. That increased battery impedance, resulting in shorter driving range.

On a side note, many car owners use a car navigation system where the destination is specified. To avoid the situation in this problem set, the fuel gauging algorithm for electric vehicles should consider the factors that shorten the driving range, such as temperature towards the destination [6].

## References


[1] Asenbauer, J., et al., "The Success Story of Graphite as a Lithium-ion Anode Material—Fundamentals, Remaining Challenges, and Recent Developments Including Silicon (Oxide) Composites," *Sustainable Energy Fuels,* Vol. 4, No. 11, 2020, pp. 5387–5416.

[2] Barsukov, Y., et al., *Battery Power Management for Portable Devices,* Norwood, MA: Artech House, 2013, pp. 158–169.

[3] How, D., et al., "State of Charge Estimation for Lithium-Ion Batteries Using Model-Based and Data-Driven Methods: A Review," *IEEE Access,* Vol. 7, 2019, pp. 136116–136136.

[4] Chemali, E., et al., "State-of-Charge Estimation of Li-ion Batteries Using Deep Neural Networks: A Machine Learning Approach," *Journal of Power Sources,* Vol. 400, 2018, pp. 242–255.

[5] Matsumura, N., et al., "Dynamic Battery Power Management Based on Battery Internal Impedance," United States Patent No. 10,684,667.

[6] Matsumura, N., et al., "Battery Life Estimation Based on Multiple Locations," United States Patent No. 10,416,241.

# 8

# FUEL CELL

## 8.1 INTRODUCTION

A fuel cell is one of the energy storage systems that converts chemical energy to electrical energy through redox reactions. Unlike Li-ion batteries, it is not rechargeable. Therefore, a fuel supply is required. The attention to fuel cells has been increasing because of their higher efficiency and lower impact to the environment, compared to combustion engines. When hydrogen is used as fuel, the fuel cell does not produce carbon dioxide, the primary greenhouse gas. It produces only water and energy. This technology has already been commercialized, especially for electric vehicles and buildings. This chapter explains how fuel cells work and their advantages/disadvantages, compared to Li-ion batteries.

## 8.2 HYDROGEN FUEL CELL

### 8.2.1 Theory

In a hydrogen fuel cell, hydrogen gas reacts with oxygen, which typically comes as part of the air. As a result, water and energy are produced. Half and overall reactions are as follows:

$$\text{Anode: } H_2 = 2H^+ + 2e^- \tag{8.1}$$

$$\text{Cathode: } 2H^+ + \tfrac{1}{2}O_2 + 2e^- = H_2O \tag{8.2}$$

$$\text{Overall: } H_2 + \tfrac{1}{2}O_2 = H_2O \tag{8.3}$$

The theoretical cell voltage is 1.23V.

Figure 8.1 is a cross-sectional image of a typical hydrogen fuel cell.

In a hydrogen fuel cell, hydrogen ions move from anode to cathode during discharge. In the anode, hydrogen comes in, and becomes hydrogen ions, $H^+$, by releasing electrons. Hydrogen ions, which are also called protons, travel to the cathode side through a proton exchange membrane. In the cathode, hydrogen ions react with oxygen in the air, and the electrons. As a result, water is formed. The overall reaction (8.3) is the opposite reaction of water split into hydrogen and oxygen.

### 8.2.2 Structure

A hydrogen fuel cell is also called PEMFC because a proton/polymer exchange membrane (PEM) is used in the fuel cell (FC). PEM is electrically insulative and hydrogen conductive. A perfluorosulfonic

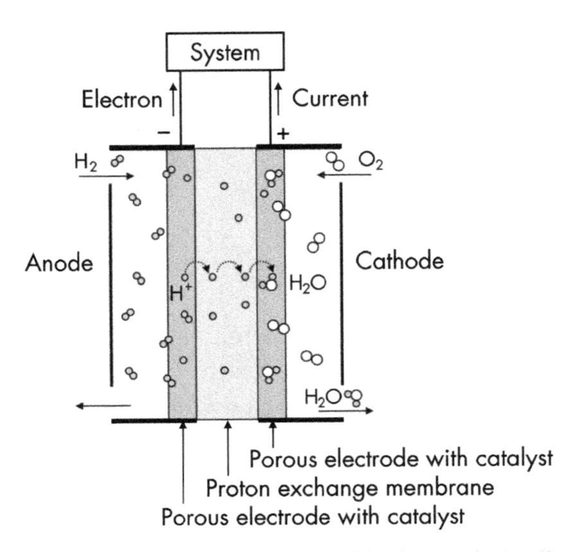


**Figure 8.1** A cross-sectional image of a typical hydrogen fuel cell.

acid, such as Nafion, is widely used as a PEM [1]. It contains the sulfonic acid group, $-SO_3H$, at the end of the chain, where hydrogen ions travel [2].

For efficient reactions, gas, ions through membrane, and electrons need to meet at the large surface area. Also, the activation energy needs to be lowered by a catalyst that is typically based on platinum. The boundary between gas, membrane, and catalytic electrode particles is called a triple phase boundary (TPB), or a three-phase boundary. To enable large TPBs, the porous electrodes (e.g., carbon) with the dispersed catalyst are used. Such porous structure also works for efficient gas diffusion. An example of TPB in the cathode side is shown in Figure 8.2.

The cathode electrode, membrane, and anode electrode are stacked and called a membrane electrode assembly (MEA).

## 8.3 FUEL CELL CHARACTERISTICS

### 8.3.1 Current Versus Voltage: I-V Curve

The theoretical cell voltage of a hydrogen fuel cell is 1.23V. However, the real OCV is lower than that. Also, the voltage decreases as the discharging current increases. For a Li-ion battery, the battery voltage drops during discharge due to IR drop, which is explained in Chapter

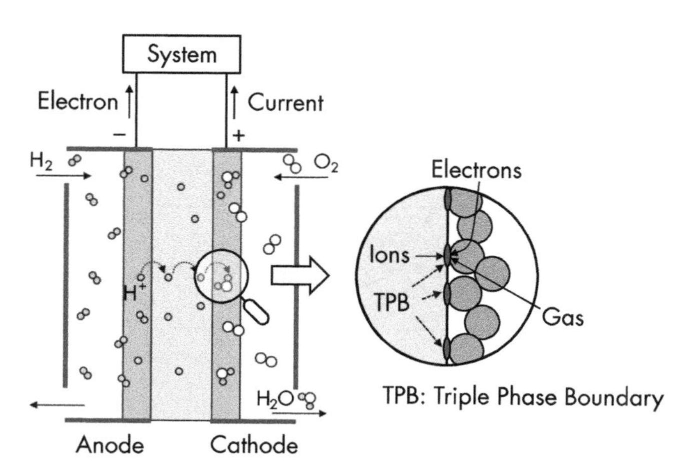


**Figure 8.2** An example of a triple phase boundary of a hydrogen fuel cell in the cathode side.

3. For a fuel cell, the voltage drop is more complex. Figure 8.3 is an example of the voltage-current relationship for a hydrogen fuel cell. The diagram is called I-V curve or I-V characteristics after current (I) and voltage (V).

As shown in the figure, the cell voltage decreases when the discharging current density increases. This is because of the four types of losses: crossover loss, activation loss, ohmic loss, and concentration loss. These losses make efficiency worse because the fuel cell provides less energy. The following explains the details of each loss in Figure 8.3.

#### 8.3.1.1 *Crossover Loss*

The OCV of a hydrogen fuel cell is lower than the theoretical voltage. This is because part of the hydrogen gas in the anode passes through the membrane into the cathode, not as ions but as gas because of its small size. This causes electrons to be released from the hydrogen in the cathode and the oxygen reduction reaction in the cathode, resulting in internal crossover current in the cathode and the crossover loss as a cell [3, 4].

#### 8.3.1.2 *Activation Loss*

When the reaction is initiated, the activation loss associated with activation energy is observed as shown in Figure 8.3. This loss is the volt-

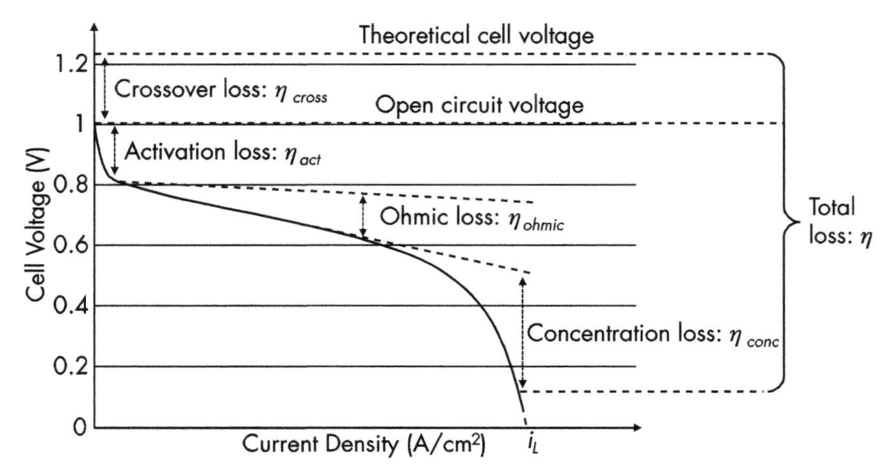


**Figure 8.3** An example of I-V curve for a hydrogen fuel cell.

age required to overcome the activation energy of the electrochemical reaction. The voltage drop of the activation loss $\eta_{act}$ is expressed as:

$$\eta_{act} = -[(RT)/(\alpha F)]\ \ln(i_0) + [(RT)/(\alpha F)]\ \ln(i) \tag{8.4}$$

$$= a + b \log i \tag{8.5}$$

where $\eta_{act}$ is the activation loss, $R$ is the gas constant (~8.314 J · mol$^{-1}$ · K$^{-1}$), $T$ is the temperature in Kelvin, $\alpha$ is the charge transfer coefficient, F is the Faraday constant (~96485 C · mol$^{-1}$), $i_0$ is the exchange current density, $i$ is the current density, and both $a$ and $b$ are constant.

Equation (8.4) is derived from the Butler-Volmer equation and the approximate equation (8.5) is called the Tafel equation. These equations explain that the activation loss is a linear function of the logarithm of the current density. This also means that the voltage drop is steep initially and less steep when current increases as shown in Figure 8.3.

#### *8.3.1.3 Ohmic Loss*

When current density increases, the ohmic loss, which is IR drop caused by the resistance, is observed in addition to the crossover loss and the activation loss. The resistance includes electrical resistance in the electron pathways, such as electrodes, and ionic resistance in the ion pathways, such as membrane. With Ohm's law, the ohmic loss $\eta_{ohmic}$ is expressed as:

$$\eta_{ohmic} = i \times R \tag{8.6}$$

where $i$ is the current density and $R$ is the resistance.

#### *8.3.1.4 Concentration Loss*

To proceed the reaction (8.3), hydrogen and oxygen fuels need to be supplied to the TPBs. Also, the formed water needs to be removed from the TPBs so that the next reaction can happen there. However, when the current density is high, fuel diffusion and water removal are delayed. Such limited mass transport causes the concentration loss in addition to the losses of crossover, activation, and ohmic. From Fick's law of diffusion and the Nernst equation, the concentration loss $\eta_{conc}$ is expressed as:

$$\eta_{conc} = [(RT)/(nF)] \ln[i_L/(i_L - i)] \tag{8.7}$$

where $R$ is the gas constant (~8.314 J · mol$^{-1}$ · K$^{-1}$), $T$ is the temperature in Kelvin, $n$ is the number of electrons, $F$ is the Faraday constant (~96485 C · mol$^{-1}$), $i_L$ the limiting current density, which is the maximum possible current density shown in Figure 8.3, and $i$ is the current density.

The total loss $\eta$ is calculated as:

$$\eta = \eta_{cross} + \eta_{act} + \eta_{ohmic} + \eta_{conc} \tag{8.8}$$

Reduction of these losses increases the efficiency of the fuel cell. For example, large electrode surface area with the dispersed catalyst reduces the activation and ohmic losses. Increase in gas pressure, fast gas supply, and fast water removal reduce the concentration loss.

### 8.3.2 Current Versus Power: I-P Curve

The power density of a fuel cell is calculated as $V \times I$ where $V$ is voltage, and $I$ is current density. Figure 8.4 is an example of a current-power relationship for a hydrogen fuel cell.

This diagram is called I-P curve or I-P characteristics after current $I$ and power $P$. The figure also shows the I-V curve for comparison. The $y$-axis of the I-P curve is on the left side and the $y$-axis of the I-V curve is on the right side. As the figure shows, the power density does not increase linearly when the current density increases. This is because the voltage decreases when the current density increases. Initially the power density increases as the current density increases. It reaches a maximum value at some point and drops when the concentration loss happens due to the limited mass transport.

### 8.3.3 Sporadic Current Change and Voltage Response

Unlike a Li-ion battery, there is no discharge cutoff voltage in a hydrogen fuel cell. Still, when the cell voltage hits the system shutdown voltage, the system shuts down. The current in real systems is typically not stable but changes frequently. Portable devices, such as laptop

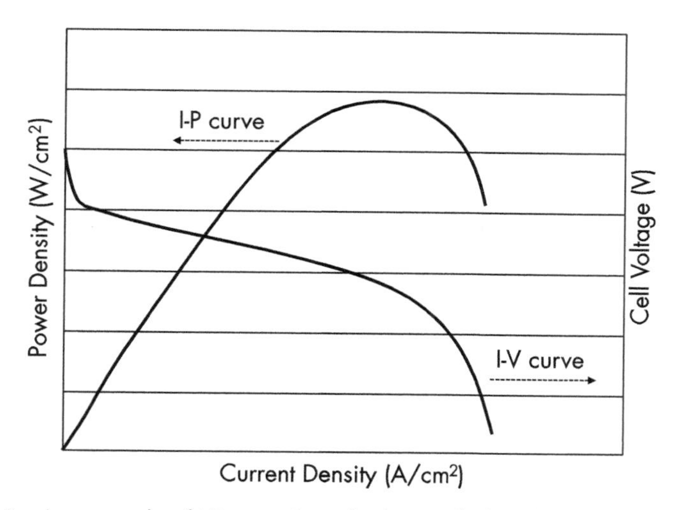


**Figure 8.4** An example of I-P curve for a hydrogen fuel cell.

PCs, perform a burst power mode that draws sporadic high current. Electric vehicles also need sporadic high current during acceleration. However, as shown in Figure 8.4, high current drops the cell voltage and may hit the system shutdown voltage. Also, there is a limit in output power. To support the high current and extend the system operation time, one option is to increase the surface area of the reaction, for example, by increasing the electrode surface area and/or stacking up multiple cells. It is also an option to make the power source a hybrid with a supplemental energy storage device that supports high current, such as a Li-ion battery.

## 8.4 TEMPERATURE AND PRESSURE IMPACTS ON PERFORMANCE

### 8.4.1 Application of Nernst Equation to Fuel Cell

For a hydrogen fuel cell, let's think about how temperature and gas pressure affect the performance. Chapter 2 explained that each substance has its own Gibbs free energy, which is defined by the following formula:

$$G = G^{\circ} + RT \ln a \tag{8.9}$$

where $G^\circ$, pronounced as G-naught, is the Gibbs free energy in standard conditions, which is at 25°C and 1 atm, $R$ is the gas constant, which is ~8.314 J · mol$^{-1}$ · K$^{-1}$, $T$ is temperature in Kelvin, $a$ is activity (e.g., gas pressure and concentration).

In reaction (8.3) of a hydrogen fuel cell, the difference of the Gibbs free energies between the right side and left side, $\Delta G$, is expressed as:

$$\Delta G = G(H_2O) - G(H_2) - ½G(O_2) \tag{8.10}$$

where $G(H_2O)$, $G(H_2)$, and $G(O_2)$ are the Gibbs free energies of $H_2O$, $H_2$, and $O_2$, respectively.

Equations (8.9) and (8.10) give:

$$\Delta G = \Delta G^\circ + RT \ln\{a(H_2O)/[a(H_2) \times a(O_2)^{1/2}]\} \tag{8.11}$$

where $a(H_2O)$, $a(H_2)$, and $a(O_2)$ are the activities (e.g., pressure) of $H_2O$, $H_2$, and $O_2$, respectively.

$\Delta G$ leads to electromotive force (emf), which is battery voltage, via the Nernst equation:

$$\Delta G = -nFE \tag{8.12}$$

where $n$ is the number of electrons transferred in the reaction, $F$ is the Faraday constant, which is ~96485 C · mol$^{-1}$, and $E$ is the theoretical cell voltage.

From (8.11) and (8.12), the theoretical cell voltage $E$, is expressed as:

$$E = \frac{1}{nF}\left(-\Delta G^\circ + RT \ln \frac{a_{H_2} a_{O_2}^{1/2}}{a_{H_2O}}\right) \tag{8.13}$$

where $\Delta G^\circ$ is –237 kJ at the standard condition for the reaction (8.3).

### 8.4.2 Pressure Impact on Voltage and Performance

Equation (8.13) means that when the pressure of hydrogen or oxygen decreases, the theoretical cell voltage decreases. Also, low gas pressure leads to early concentration loss even at low current density. As a result, maximum power density decreases when the gas pressure decreases. Figure 8.5 shows the I-V and I-P curves in case of pure oxygen and the air, which contains 21% oxygen.

As shown in Figure 8.5(a, b), the air contains lower pressure of oxygen than the pure oxygen, so both voltage and the maximum output power are lower, compared to the pure oxygen. Figure 8.5(a) also shows that the voltage drop due to concentration loss starts earlier with air than with pure oxygen.

### 8.4.3 Temperature Impact on Voltage

Temperature affects the $\Delta G^{\circ}$ and the theoretical voltage in (8.13). However, the theoretical voltage change by temperature at open circuit is very small, compared to the voltage-drop change by temperature during discharge. When temperature increases, voltage drop decreases, resulting in higher operating voltage and higher power density [5]. This is because, at the elevated temperature, resistance in the membrane is lowered and the reactions on the catalyst in the electrode are accelerated.

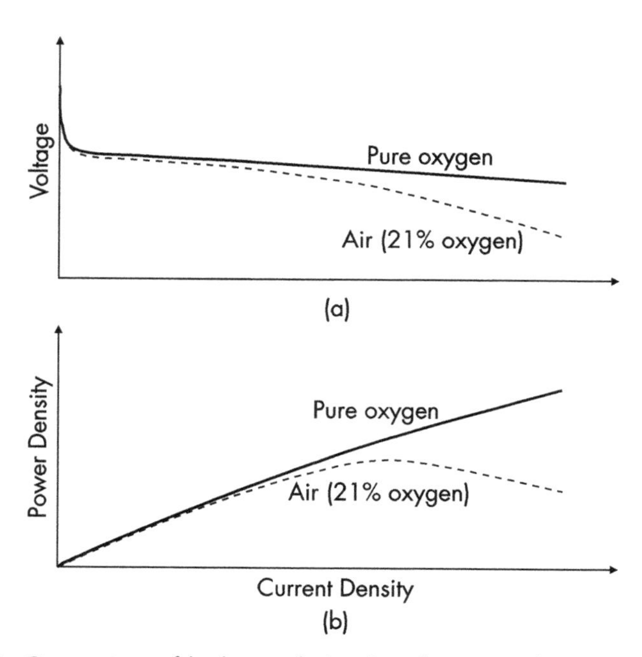


**Figure 8.5** Comparison of hydrogen fuel cell performances between pure oxygen and air (21% oxygen): (a) I-V curves, and (b) I-P curves.

## 8.5 OTHER FUEL CELLS

There are many other fuel cells that use different fuels and membranes. This section introduces two commercialized fuel cells, a direct methanol fuel cell (DMFC) and a solid oxide fuel cell (SOFC).

### 8.5.1 Direct Methanol Fuel Cell

In a DMFC, a methanol solution ($CH_3OH + H_2O$) is directly supplied to the anode. Cathode fuel is oxygen or the air that contains oxygen. The following are the half and overall reactions:

$$\text{Anode: } CH_3OH + H_2O = 6H^+ + CO_2 + 6e^- \tag{8.14}$$

$$\text{Cathode: } 3/2O_2 + 6H^+ + 6e^- = 3H_2O \tag{8.15}$$

$$\text{Overall: } CH_3OH + 3/2O_2 = CO_2 + 2H_2O \tag{8.16}$$

Figure 8.6 is an example of a direct methanol fuel cell.

At the anode side, hydrogen ions and carbon dioxide are formed, releasing electrons. The hydrogen ions move through the PEM that is

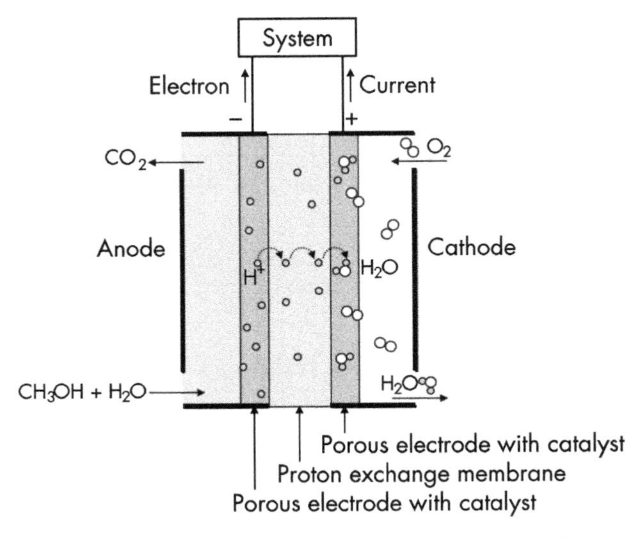


**Figure 8.6** A cross-sectional image of a direct methanol fuel cell.

also used in a hydrogen fuel cell, and react with oxygen at the cathode side, forming water.

The DMFC is a safer PEMFC than the hydrogen fuel cell because the methanol solution is easier to carry/handle than the high-pressure hydrogen gas. Although the DMFC was commercialized, such as for a charger of portable mobile devices, it was not commercially successful. This is partly because of the difficulty in the toxic methanol supply chain to the end users, and the competition against the more convenient and affordable chargers with Li-ion batteries.

### 8.5.2 Solid Oxide Fuel Cell

In an SOFC, hydrogen or methane ($CH_4$) is typically used for anode, and oxygen or the air containing oxygen is used for cathode. In SOFC, oxygen ions move from cathode to anode through an oxygen-ion conductive solid-state electrolyte. The half and overall reactions are as follows.

In the case of hydrogen:

$$\text{Cathode: } ½O_2 + 2e^- = O^{2-} \tag{8.17}$$

$$\text{Anode: } H_2 + O^{2-} = H_2O + 2e^- \tag{8.18}$$

$$\text{Overall: } H_2 + ½O_2 = H_2O \tag{8.19}$$

In the case of methane:

Cathode: The same with reaction 8.17

$$\text{Anode: } CH_4 + 4O^{2-} = 2H_2O + CO_2 + 8e^- \tag{8.20}$$

$$\text{Overall: } CH_4 + 2O_2 = 2H_2O + CO_2 \tag{8.21}$$

Figure 8.7 is an example of a solid oxide fuel cell.

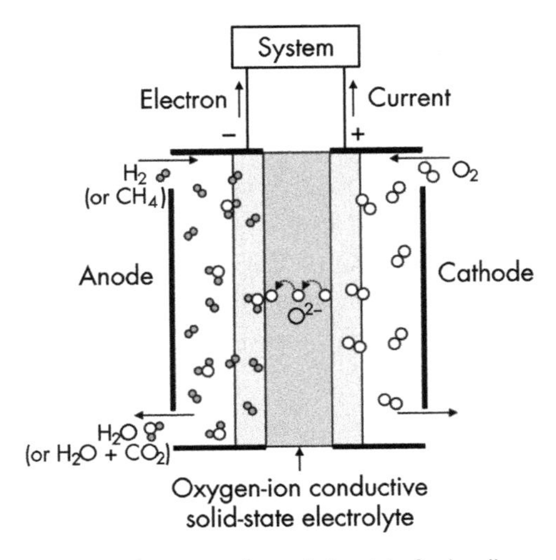


**Figure 8.7** A cross-sectional image of a solid oxide fuel cell.

The oxygen-ion conductive solid-state electrolyte is a ceramic that has vacancies in the crystal structure, for example, yttria-stabilized zirconia (YSZ). The oxygen ions formed in the cathode jump through the vacancies in the solid-state electrolyte into the anode side and react with the gas in the anode. SOFC is operated at high temperatures, such as 500°C to 1000°C [6]. Such high temperature increases the conductivity of the solid-state electrolyte and enhances efficiency.

When the anode gas is hydrogen, water is formed via (8.18). When methane is used as the anode gas, methane, precisely speaking the decomposed methane such as hydrogen, reacts with oxygen ions and forms water and carbon dioxide as expressed in (8.20). The methane decomposition is accelerated at high temperatures [7].

While it takes time and energy to start the SOFC operation due to the high operating temperature, because of its higher efficiency than that of the conventional combustion-based power plants [8], SOFC is commercialized by many companies. SOFC can use not only hydrogen but also methane, which is the largest component of natural gas, such as ~93%. Natural gas is already supplied to many houses and buildings, which already addresses the commercial challenges of fuel supply.

## 8.6 FUEL CELLS COMPARISON TO LI-ION BATTERY

Table 8.1 shows the comparison between PEMFC (with hydrogen), SOFC, and Li-ion battery.

Availability of fuel, recharging method, and operating temperature are part of the factors that determine the applications. For example, PEMFC (hydrogen) operates at the temperature where humans typically live, except in the cold environments. One of the applications is electric vehicles (fuel-cell vehicles) at the place where hydrogen refueling stations are available. SOFC operates with natural gas, which contains methane, and a stationary power supply is an example of the applications. Li-ion batteries are rechargeable at the place where electricity is available. While charging takes longer than hydrogen refueling, as there are many power outlets and charging stations, it is widely used for portable devices and electric vehicles.

PEMFC (hydrogen) and SOFC systems generate electricity from a fuel at efficiencies up to 60% [8], whereas Li-ion batteries output the charged energy at efficiencies more than 95% depending on discharging conditions. For the details of losses for Li-ion batteries, refer to Chapter 3. In addition to these efficiencies, the losses in the process of creating a fuel and delivering power in a system need to be considered to compare the true efficiencies. For example, in the case of hydrogen fuel-cell vehicles versus battery electric vehicles, overall efficiencies are 23% versus 76%, respectively [9]. The difference is mainly attributed to consumed energies at hydrogen production by

**Table 8.1**
An Example of Comparison Between PEMFC (Hydrogen), SOFC, and Li-ion Battery

| Technology | Charge Carrier | Recharging Method | Operating Temperature | Requirements | Applications |
|---|---|---|---|---|---|
| PEMFC (Hydrogen) | $H^+$ | Refueling | ≥ 0°C to start < 120°C [6] | Anode tank (hydrogen) | Electric vehicles |
| SOFC | $O^{2-}$ | Refueling | 500°C ~ 1000°C | Anode gas supply (hydrogen or methane) | Stationary power supplies |
| Li-ion battery | $Li^+$ | Plug-in | −20°C ~ 60°C | Long charging time, anode and cathode materials | Portable devices, electric vehicles |

electrolysis and hydrogen transportation, in addition to efficiency difference between a PEMFC (hydrogen) and a Li-ion battery. It is also important to consider total cost of ownership which, for example, includes the initial purchase price of the vehicle and the cost of operation, such as fuel/electricity cost, cell/battery replacement cost, and the resale value.

## 8.7 FUEL CELL EXPERIMENTS WITH A HYDROGEN FUEL-CELL KIT

In the market (e.g., online stores), several hydrogen fuel-cell kits are available for educational purposes. While these kits are typically designed for middle school students, they are also a great resource to confirm the knowledge covered in this chapter. This section explains some experimental procedures with the kits. Figure 8.8 is an example of a hydrogen fuel cell kit.

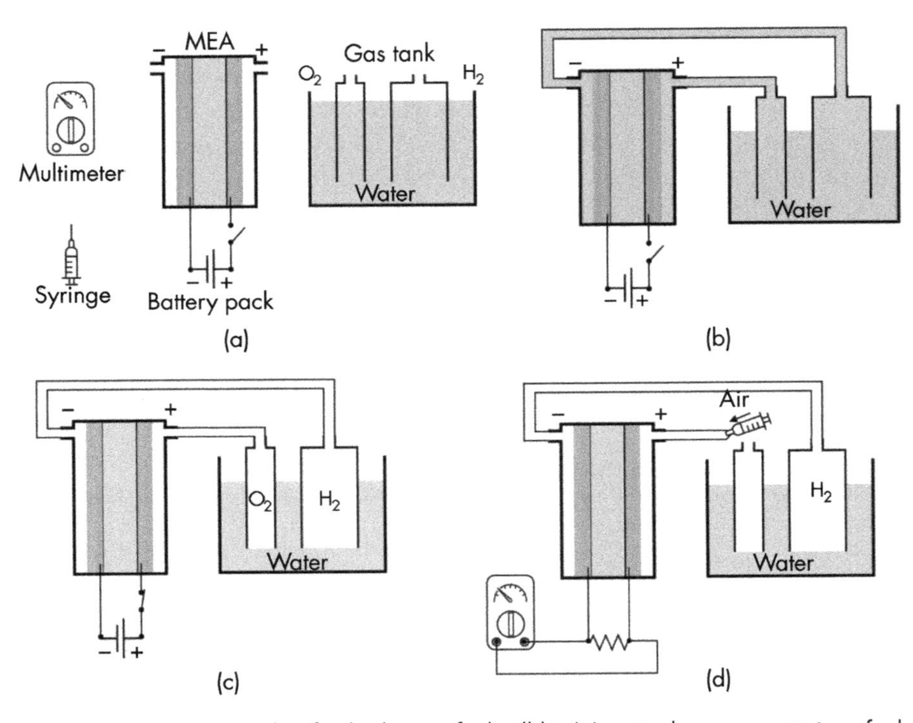


**Figure 8.8** An example of a hydrogen fuel-cell kit (a) typical components in a fuel-cell kit, (b) after the MEA and the gas tank are filled with water, (c) after water is split into hydrogen and oxygen, and (d) when oxygen is replaced with the air.

The kit typically includes a fuel cell (i.e., MEA), a gas tank, a syringe, a battery pack, and a multimeter as shown in Figure 8.8(a). A hydrogen fuel cell works with hydrogen and oxygen. In the kit, these gases are produced by splitting water with the battery pack. To do that, first, the MEA and the gas tank are filled with water as shown in Figure 8.8(b). Then, the water in the MEA is split with the battery pack and the generated hydrogen and oxygen are sent to each tank as shown in Figure 8.8(c). The following are some examples of experiments and the expected results.

Disclaimer: Use of the kits and these examples is at your own risk.

*OCV:* Measure the fuel-cell voltage with the multimeter after splitting water and disconnecting the battery pack. The observed voltage is lower than the theoretical voltage 1.23V. This is because of the crossover loss.

*Voltage during operation:* Instead of the battery pack, connect several types of resistors to the fuel cell for discharging. Measure the fuel-cell voltage during discharge with the multimeter and calculate the current by $V/R$ where $V$ is the measured voltage, and $R$ is the resistance of the resistor. If a battery tester is available, it may be connected to the fuel cell instead of the resistors, and the constant current discharging may be performed at several current levels while measuring the fuel-cell voltage. The I-V curve with pure oxygen can be drawn with the data. You will see that the observed voltage is even lower than OCV. This is because of additional losses. Depending on the current, activation loss, ohmic loss, and concentration loss happen.

*Voltage during operation with the air instead of pure oxygen:* After splitting water, replace the oxygen connection with the syringe as shown in Figure 8.8(d) and continuously supply the air with the syringe. Connect several types of resistors to the fuel cell as performed in the previous experiment, measure the fuel cell voltage, and calculate the current. If a battery tester is available, connect it to the fuel cell instead of the resistors, and conduct constant current discharging at several current levels while measuring the fuel-cell voltage. The I-V curve with the air can be drawn with the data. The voltage of the curve with the air is lower than that with pure oxygen. This is because the activity (i.e., pressure) of oxygen in the air is lower than that of pure oxygen.

## 8.8 SUMMARY

In this chapter, we learned the following:

- How fuel cells work.
- Four types of losses (crossover, activation, ohmic, and concentration).
- The comparison between fuel cells and a Li-ion battery.

## 8.9 PROBLEMS

There are frequent questions about energy density for a fuel cell and a Li-ion battery. It highly depends on the size of the system and application. This section explains a simple example to calculate energy density, which can be extended to different use cases.

Problem 8.1

There is an electric vehicle that runs on a hydrogen fuel cell. The fuel-cell system consists of the fuel-cell module, including stacked MEAs, and the high-pressure hydrogen tank. The fuel cell uses hydrogen from the tank and oxygen in the air. The tank contains 5.6 kg of hydrogen at the maximum pressure. The volume of the tank and the fuel-cell module are 123 liters and 37 liters, respectively. The molar mass of $H_2$ is 2.0 g $\cdot$ mol$^{-1}$. The Faraday constant is ~96485 C $\cdot$ mol$^{-1}$. When the fuel cell operates at 0.70V on average, answer the following problems.

1. When all hydrogen in the tank is used, how many watt-hours can the fuel-cell system theoretically provide? Ignore the fuel crossover and the hydrogen remaining in the tank.

2. When the required components for the fuel-cell system are only the hydrogen tank and the fuel-cell module, what is the volumetric energy density of the fuel-cell system?

3. When the car drives 4.0 miles per kWh, how many miles can the car theoretically drive with the full hydrogen tank? Ignore loss through power delivery, such as losses by impedance and voltage regulation.

4. When hydrogen cost is $14/kg, how much is the refueling cost from empty?

5. If the same watt-hours are provided from a Li-ion battery instead of the hydrogen fuel cell, and electricity cost is \$0.10/kWh, how much is the recharging cost from empty?

Answer 8.1

1. A 5.6 kg of hydrogen is equivalent to $5.6 \text{ kg} \times 1000/2.0\text{g} \cdot \text{mol}^{-1} = 2800$ mol. $H_2$ releases two electrons as shown in (8.1). This means that $2800 \times 2 = 5600$ mol of electrons are released. With the Faraday constant, this is equivalent to $5600 \text{ mol} \times 96485\text{C} \cdot \text{mol}^{-1} \approx 5.4 \times 10^8\text{C}$. As $1 \text{ Ah} = 1\text{A} \times 1\text{h} = 1 \text{ C} \cdot \text{s}^{-1} \times 3600\text{s} = 3600\text{C}$, $5.4 \times 10^8\text{C}$ equals $5.4 \times 10^8\text{C}/3600 \text{ C} \cdot \text{Ah}^{-1} \approx 1.5 \times 10^5$ Ah. The average voltage is 0.70V. Therefore, theoretical energy is $1.5 \times 10^5 \text{ Ah} \times 0.70\text{V} = 1.05 \times 10^5 \text{ Wh} \approx 1.1 \times 10^5$ Wh.

2. The total volume of the system is $123\text{L} + 37\text{L} = 160\text{L}$. The volumetric energy density is $1.5 \times 10^5 \text{ Ah} \times 0.70\text{V} / 160\text{L} \approx 6.6 \times 10^2$ Wh/L.

3. $4.0 \text{ miles/kWh} \times 1.5 \times 10^5 \text{ Ah} \times 0.70\text{V} /1000 = 4.2 \times 10^2$ miles is the theoretical driving range.

4. The tank contains 5.6 kg of hydrogen at maximum. When hydrogen costs \$14/kg, the cost of refueling from empty is 5.6 kg × \$14/kg = \$78.4.

5. When the energy is provided from a Li-ion battery and the electricity cost is \$0.10/kWh, recharging from empty to full costs $1.5 \times 10^5 \text{ Ah} \times 0.70\text{V} \times \$0.10/\text{kWh}/1000 = \$10.5$.

On a side note, the energy density of a fuel cell highly depends on the size of the tank, the size of the fuel-cell module, and gas pressure. For example, the larger the tank is and/or the higher the hydrogen pressure is, the higher energy density is expected. In addition, the size and efficiency of the other components, such as voltage regulators and gas-flow systems, need to be considered, which decreases energy density. For reference, while the energy density of some Li-ion battery cells exceeds 700 Wh/L, a Li-ion battery pack needs not only cells but also a battery management system for safety, and the space between the cells for thermal management. These decrease energy density. In the problem sets, refueling and recharging costs are also calculated for a hydrogen fuel cell and a Li-ion battery, respectively. While refueling hydrogen costs more, it completes in several minutes, which is faster than recharging a Li-ion battery. For cost comparison, it is also important to consider the initial cost and resale value of the vehicles.

## References

[1] Feng, C., et al., "Mechanical Behavior of a Hydrated Perfluorosulfonic Acid Membrane at Meso and Nano Scales," *RSC Adv.,* Vol. 9, 2019, pp. 9594–9603.

[2] Li, T., et al., "Performance Comparison of Proton Exchange Membrane Fuel Cells with Nafion and Aquivion Perfluorosulfonic Acids with Different Equivalent Weights as the Electrode Binders," *ACS Omega,* Vol. 5, No. 28, 2020, pp. 17628–17636.

[3] Vilekar, S., et al., "The Effect of Hydrogen Crossover on Open-Circuit Voltage in Polymer Electrolyte Membrane Fuel Cells," *Journal of Power Sources,* Vol. 195, No. 8, 2010, pp. 2241–2247.

[4] Li, S., et al., "Voltammetric and Galvanostatic Methods for Measuring Hydrogen Crossover in Fuel Cell," *iScience,* Vol. 25, No. 34984330, 2021, p. 103576.

[5] Santarelli, M. G., et al., "Experimental Analysis of the Effects of the Operating Variables on the Performance of a Single PEMFC," *Energy Conversion and Management,* Vol. 48, No. 1, 2007, pp. 40–51.

[6] U.S. Department of Energy, Fuel Cell Technologies Office, "Comparison of Fuel Cell Technologies," https://www.energy.gov/sites/default/files/2016/06/f32/fcto_fuel_cells_comparison_chart_apr2016.pdf.

[7] Ideris, A., et al., "Direct-Methane Solid Oxide Fuel Cell (SOFC) with Ni-SDC Anode-Supported Cell," *International Journal of Hydrogen Energy,* Vol. 42, No. 36, 2017, pp. 23118–23129.

[8] U.S. Department of Energy, https://www.energy.gov/sites/prod/files/2015/11/f27/fcto_fuel_cells_fact_sheet.pdf.

[9] Tsakiris, A., "Analysis of Hydrogen Fuel Cell and Battery Efficiency," *World Sustainable Energy Days 2019,* Wels, Austria, Feb. 27–Mar. 1, 2019.

# 9

# OTHER BATTERY-RELATED TECHNOLOGIES

## 9.1 INTRODUCTION

This chapter explains the basics of the technologies that are related to batteries. While these technologies are not batteries, understanding the technologies will open the door for creative system designs and better collaborations with the experts in those areas.

## 9.2 SUPERCAPACITORS

### 9.2.1 Theory

A supercapacitor is an electrochemical capacitor that stores energy in its electric double layers. It is also known as an ultracapacitor. Figure 9.1 is the comparison of a supercapacitor and a Li-ion battery during charge.

In the supercapacitor shown in Figure 9.1(a), when voltage is applied, the positive ions in the electrolyte, such as $K^+$ in KOH aqueous electrolyte [1], move towards the negative electrode (carbon) that is negatively charged. At the same time, the negative ions, such as $OH^-$ in KOH aqueous electrolyte [1], move towards the positive electrode (carbon) that is positively charged. As a result, electric double layers are formed at the boundary between electrolyte and electrodes. These

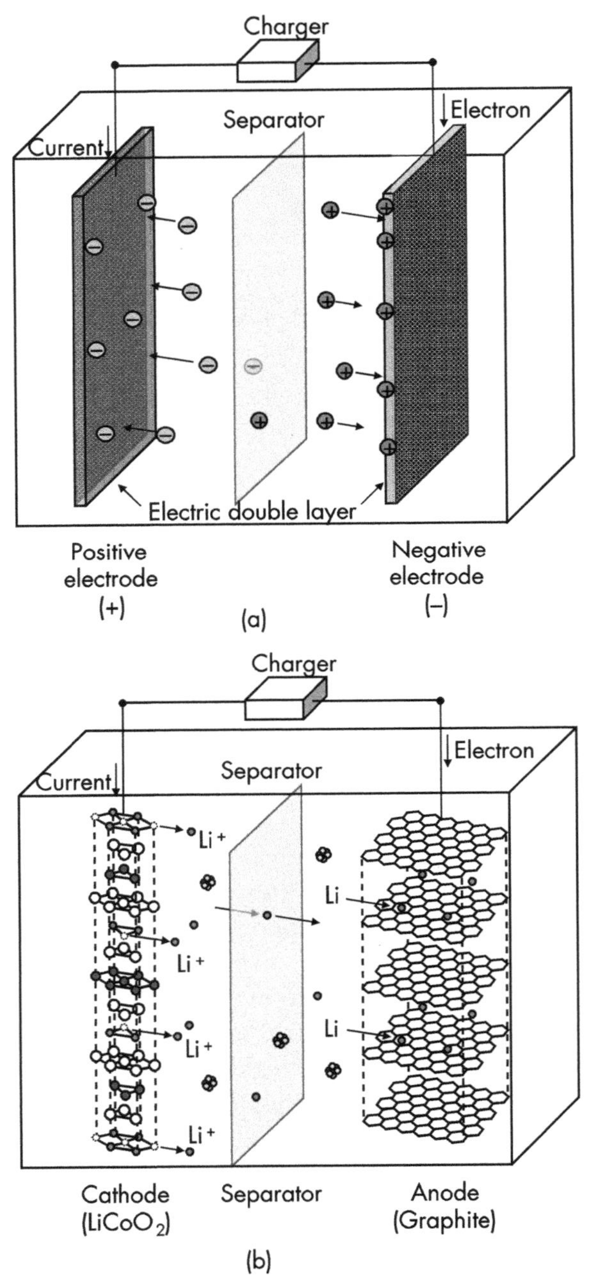


**Figure 9.1** Comparison of (a) a supercapacitor, and (b) a Li-ion battery during charge.

electric charges in the electric double layers work as energy storage. Such a supercapacitor is called an electric double layer capacitor (EDLC). While an EDLC stores energy by charges, a Li-ion battery stores energy by electrochemically moving lithium ions from cathode to anode and storing them at the different level of the Gibbs free energy, as shown in Figure 9.1(b).

Some supercapacitors have polymers or transition metal oxide in the electrode, such as $MnO_2$ [2]. When voltage is applied, the number of electrons in the transition metal changes, such as from Mn(IV) to Mn(III), and receives cations in the electrolyte in addition to electric double layers. Such a capacitor is called a pseudocapacitor.

The capacity of the supercapacitor, $Q$, is expressed as:

$$Q = CV \tag{9.1}$$

where $C$ is the capacitance of the capacitor in Farad, and $V$ is the voltage of the capacitor that needs to be below the decomposition voltage of the electrolyte. The capacitance and the applicable voltage are written in the spec of the capacitor.

The larger the electrode surface area is, the more ions are attracted to the electrodes, resulting in higher capacity. Therefore, materials that have large surface area, such as activated carbon that has etched surfaces (e.g., 1670 $m^2/g$), nano-sized carbon, graphite, and graphene, are typically used.

### 9.2.2 Structure

Figure 9.2 is an example of a supercapacitor structure.

The structure is similar to a Li-ion battery. A supercapacitor consists of two electrodes and a separator. The electrodes consist of the electrode material and current collectors. Both sides of the current collectors are coated with the electrode material, such as activated carbon. The separator is placed in between the electrodes to avoid internal short circuit. The separator is electrically insulative but ion-permeable so that the ions can go through it during charge and discharge.

The electrodes and the separators are wound together. The cell assembly is inserted into a case with an electrolyte to form a cylindrical

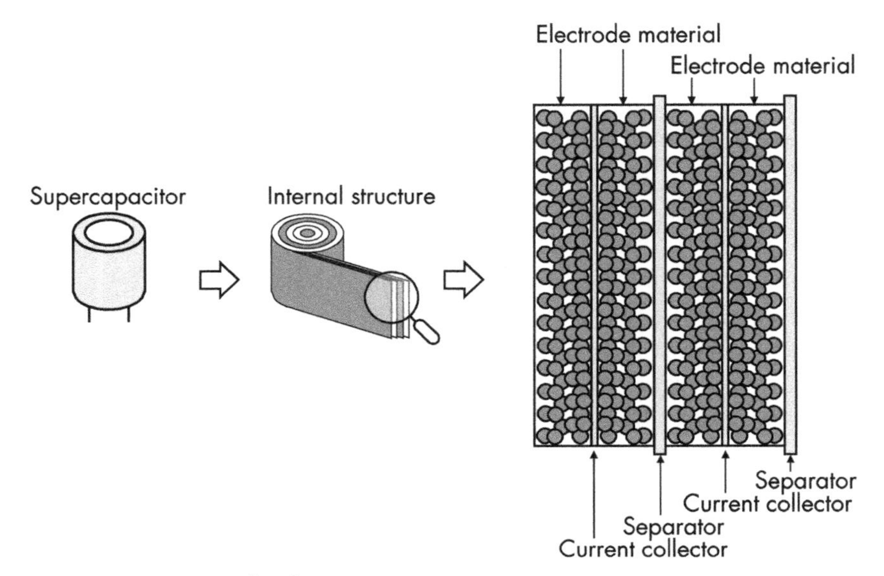


**Figure 9.2** An example of a supercapacitor structure.

cell with the winding structure. A rectangular-shaped supercapacitor is also possible.

### 9.2.3 Advantages and Disadvantages

As the electrodes of a supercapacitor are loosely attracting the ions, there is almost no chemical reaction for charging and discharging. This leads to longer cycle life (e.g., several hundred-thousand cycles) than that of a Li-ion battery (e.g., several hundred cycles). Such loosely attracted ions can be released fast, which leads to higher power output and faster charging with lower impedance, compared to a Li-ion battery. For example, a supercapacitor provides several times higher power than a Li-ion battery with the same size.

However, in a supercapacitor, after charging, the ions in the electrolyte do not strongly react with the electrode materials, unlike a Li-ion battery. This leads to the following disadvantages.

Self-discharge rate within a supercapacitor is high. For example, more than 20% of capacity is self-discharged in a month, whereas the self-discharge rate of a fully charged Li-ion battery may be several percent in a month.

Energy density of a supercapacitor is lower than that of a Li-ion battery. For example, supercapacitor is several to ten Wh/l, whereas a Li-ion battery is 700 Wh/l.

The voltage of a supercapacitor is lower than that of a Li-ion battery and goes to zero voltages. Boost circuitry may be needed for the supercapacitor to support certain voltage rails of a system operation.

Because of these advantages and disadvantages, supercapacitors are typically used not as a primary high energy-density storage, but as a supplemental energy storage for high-power output and fast charging. For example, for electric vehicles, the recuperated energy by the sporadic brakes can be stored in supercapacitors because it is capable of fast charging. Some systems may draw low current from the high energy-density but high-impedance battery, such as a nonrechargeable lithium thionyl chloride battery ($LiSOCl_2$). If such systems need to draw sporadic high current, the systems may also include supercapacitors to support the high discharging current. Because of the low impedance of a supercapacitor, hitting the system shutdown voltage is delayed and the battery life can be extended.

There are several methods to make a hybrid system with a Li-ion battery and a supercapacitor, such as passive and active configurations [3]. In passive configuration, a Li-ion battery and a supercapacitor are directly connected in parallel. While this is an affordable design, the voltages of a battery and supercapacitor are leveled, which limits each capability. Also, the self-discharge rate in a supercapacitor may consume the energy in a Li-ion battery over time. In active configuration, a supercapacitor and/or a Li-ion battery are actively controlled at the expense of the cost and space of the controlling system. For example, a supercapacitor can take the recuperated energy by the sporadic brakes of an electric vehicle before a Li-ion battery and release it for the next discharge before discharging the Li-ion battery. When a supercapacitor becomes empty, a Li-ion battery can take over the energy supply to the vehicle but does not supply the energy to the supercapacitor. This prevents self-discharge in a supercapacitor from taking energy in a Li-ion battery.

It is also important to assess the true benefit of a hybrid system. Adding a supercapacitor to a Li-ion battery takes additional space and weight. In some cases, this still benefits the system performance. In other cases, utilizing the additional space to make the Li-ion battery larger and/or thicker without the supercapacitor may be beneficial because the larger and/or thicker Li-ion battery provides higher capacity with lower impedance.

### 9.2.4 Energy Calculation

Figure 9.3 shows the discharge curves of a supercapacitor and a Li-ion battery.

Energy of a supercapacitor, $E$, can be expressed as:

$$E = \int_0^{Capacity} V\,dQ \tag{9.2}$$

$$= (1/2)CV^2 \tag{9.3}$$

where $Q$ is the capacity that equals $CV$ as shown in (9.1), $C$ is the capacitance of the supercapacitor, and $V$ is the rated voltage or the charged voltage.

This is equivalent to the shaded area of OVQ in Figure 9.3. The unit of capacitor energy, $E$, is Joule (J).

For a reference, as explained in Chapter 3, energy of a Li-ion battery, $E_{Li\_ion}$, is expressed as:

$$E_{Li\_ion} = \int_0^{Capacity} V\,dAh \tag{9.4}$$

where $V$ is the battery voltage, and $Ah$ is the battery capacity.

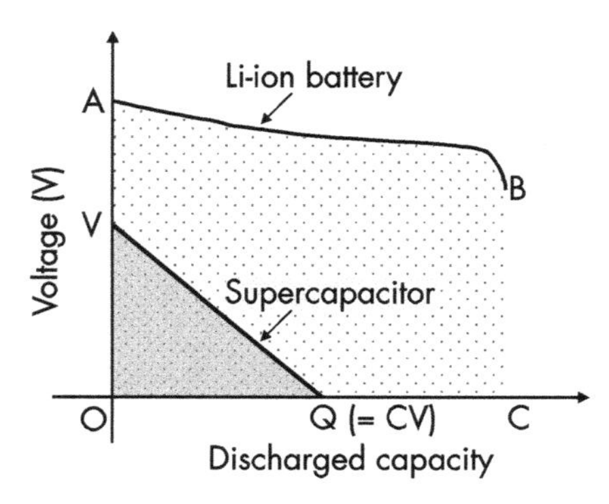


**Figure 9.3** Schematic illustration of discharge curves for a supercapacitor and a Li-ion battery.

This is equivalent to the dotted area of OABC in Figure 9.3. The unit of Li-ion battery energy, $E_{Li\_ion}$, is Wh.

Both Joule and Wh are the units for energy. As Wh is W × hour = (J/sec) × 3600 sec = 3600J, Joule and Wh can be converted to each other via the following equation:

$$1 \text{ Wh} = 3600\text{J} \tag{9.5}$$

***Exercise***

When the supercapacitor with 0.4F capacitance is charged at 2.5V, what is the stored energy in Joule and in Wh?

***Answer***

$\frac{1}{2} \times 0.4\text{F} \times (2.5\text{V})^2 = 1.25\text{J}$. This is equivalent to $1.25\text{J} / 3600 \approx 0.35$ mWh.

* * *

On a side note, IR drop happens during discharge. Therefore, the actual energy decreases depending on the loss by IR.

### 9.2.5 Li-ion Capacitor

A Li-ion capacitor consists of one Li-ion electrode and one supercapacitor electrode, for example, a Li-ion battery anode (graphite), and a supercapacitor cathode (activated carbon). Figure 9.4 is an example of a Li-ion capacitor during charge.

As a Li-ion capacitor is a hybrid of a Li-ion battery and a supercapacitor, its characteristics are typically in between a Li-ion battery and a supercapacitor. Compared to a supercapacitor, the self-discharge rate is smaller, and the operating voltage is higher. Compared to a Li-ion battery, the power capability is higher, the cycle life is longer, but the operating voltage is lower. For example, the charge cutoff voltage of a Li-ion capacitor is 3.8V and discharge cutoff voltage is 2.2V. The energy density of a Li-ion capacitor is higher than that of a supercapacitor but is lower than that of a Li-ion battery.

Figure 9.5 shows an example of the discharge curves comparison between a Li-ion capacitor, a supercapacitor, and a Li-ion battery.

The dotted area of OA′B′C′ corresponds to the discharged energy of the Li-ion capacitor.

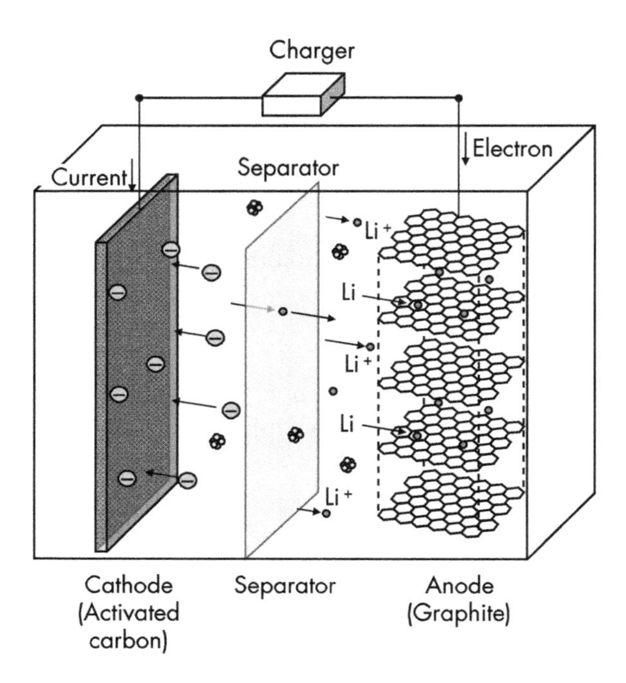


**Figure 9.4** An example of a Li-ion capacitor during charge.

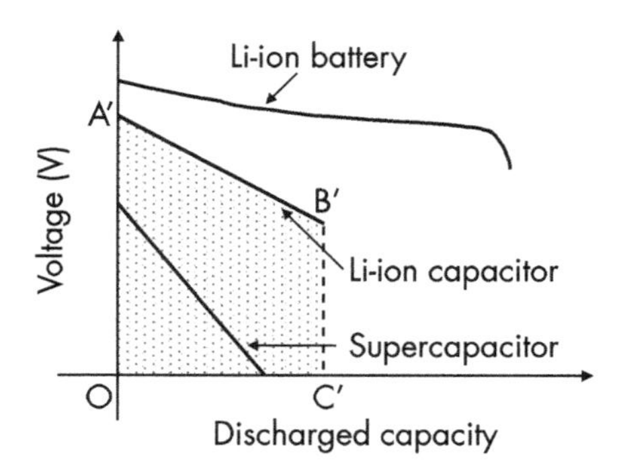


**Figure 9.5** Schematic illustration of discharge curves for a Li-ion capacitor, a supercapacitor, and a Li-ion battery.

Considering the characteristics, Li-ion capacitors are suitable for systems that need higher energy density than capacitors, and longer cycle life than Li-ion batteries. For example, Li-ion capacitors are used as a supplemental power source to energy-harvesting devices such as a solar cell.

# 9.3 SOLAR CELL

## 9.3.1 Introduction

A solar cell, also called a photovoltaic cell, is a device that converts light energy into electricity by the photovoltaic effect. As we receive a lot of the radiation energy from the sun, solar cells are widely used as solar panels on the roofs of the houses and buildings, and on the space station and satellites. Among many types of solar cells, a silicon solar cell is typically used. This section explains the theory of a silicon solar cell, unless stated otherwise, and its application to the systems for battery life extension.

## 9.3.2 Total Energy from the Sun and Efficiency of a Commercial Solar Cell

At the outer atmosphere of the Earth, 1366.1 W/m$^2$ of the integrated spectral irradiance arrives [4]. Figure 9.6 shows the relationship between spectral irradiance and wavelength [5].

The figure shows not only the visible wavelength, but also the invisible wavelength which consists of ultraviolet radiation (UV) and infrared radiation (IR). There are two lines in the figure, AM0 and AM1.5.

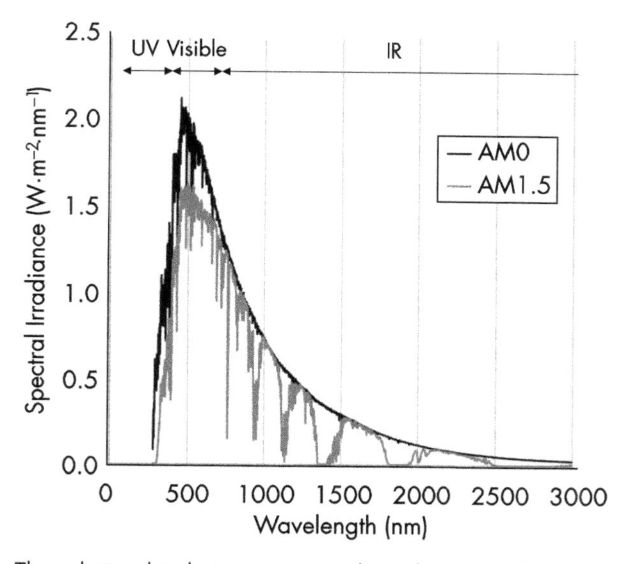


**Figure 9.6** The relationship between spectral irradiance and wavelength. (Spectral irradiance and wavelength data is after [5].)

AM is the abbreviation of air mass, and the number after AM is defined by $L/L_0$, where $L$ is the path length of the light though the atmosphere, and $L_0$ is the light path length through the atmosphere at zenith, which is directly above the measurement position. Figure 9.7 shows the definitions of AM0, AM1.0, and AM1.5.

AM0 is outside of the atmosphere, where 1366.1 W/m$^2$ of the integrated spectral irradiance arrives. Before the light goes through the atmosphere, the path length $L$ is zero. Therefore, it is called AM0. AM1.0 is the vertical illumination to the sea level through the atmosphere. As $L$ and $L_0$ are the same and $L/L_0$ is 1, it is called AM1.0. AM1.5 is the radiation to the sea level from an angle of 48.2 degrees.

Looking back at Figure 9.6, the spectral irradiance of AM1.5 is reduced from AM0 because of scattering/diffraction through the atmosphere and absorption at the certain wavelengths by the substances in the atmosphere, such as water vapor, ozone, and oxygen. The integrated spectral irradiance based on AM1.5 is ~1000 W/m$^2$. AM1.5 is commonly used in the solar cell industry as it covers the overall yearly average of midlatitudes where there is a large population.

While 1000 W/m$^2$ on average comes to solar cells, the output power from the solar cells is less than 1000 W/m$^2$ because of the solar cell efficiency. Solar cell efficiency, $\eta$, is calculated as:

$$\eta = P_{out}/P_{in} \tag{9.6}$$

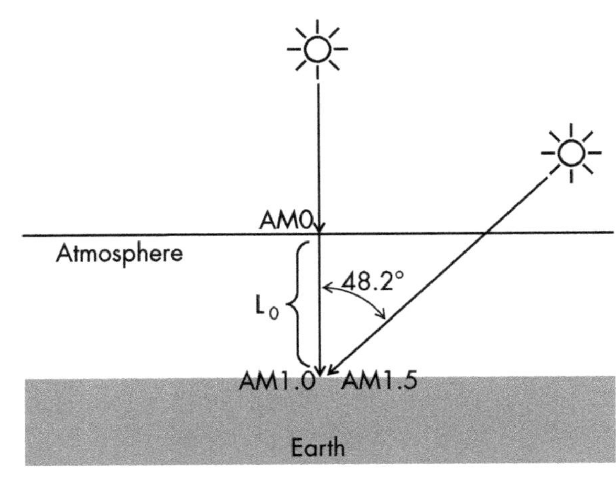


**Figure 9.7** Definitions of AM0, AM1.0, and AM1.5.

where $P_{out}$ is the power delivered out of the solar cell and $P_{in}$ is the input power to the solar cell, which is 1000 W/m$^2$ (i.e., 0.1 W/cm$^2$) in case of AM1.5.

In the 1950s, the efficiency of the commercialized solar cell was 10% or less. This means that if there is a solar cell that is 1m × 1m and has 10% efficiency, the output based on AM1.5 is 1000 W/m$^2$ × 1m × 1m × 10% = 100W.

Thanks to the efforts of researchers and engineers, the efficiency of the commercially available solar cells for roofs has increased to ~20%. This means that the output of the 1m × 1m solar cell is ~200W.

### 9.3.3 Theory

A silicon solar cell consists of n-type semiconductor and p-type semiconductor. Figure 9.8 is a schematic illustration of n-type and p-type semiconductors.

Silicon (Si) belongs to the group 14 elements that have four valence electrons. When a group 15 element, such as phosphorus (P), is doped in silicon, the silicon has one more electron because the group 15 element has one more valence electron than silicon. As a result, the silicon has one free electron as shown in Figure 9.8(a). As the free electron has a negative charge, it is called n-type. For a p-type semiconductor, a group 13 element, such as boron (B), is doped on silicon. This creates one less valence electron because the group 13 element has one less valence electron than silicon. Such silicon creates a hole that has a positive charge as shown in Figure 9.8(b) and is called p-type.

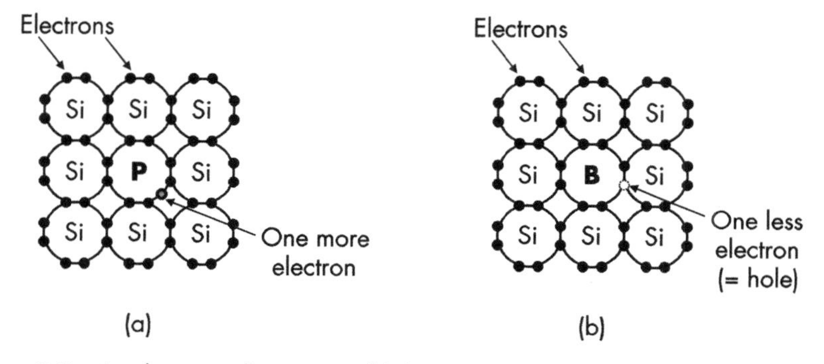


**Figure 9.8** A schematic illustration of (a) n-type semiconductor, and (b) p-type semiconductor.

When n-type and p-type semiconductors are joined together as shown in Figure 9.9, the electrons from the n-type semiconductor diffuse across the junction and are compensated with the holes in the p-type semiconductor. Then, the region of the p-type semiconductor near the junction is negatively charged because the electrons are attracted by the p-type semiconductor. Similarly, the region of the n-type semiconductor near the junction is positively charged. Such a positively and negatively charged region near the junction is called the depletion region.

The p-n junction is the basic structure of a solar cell. When the p-n junction is illuminated and the photon energy is greater than band gap energy between the valence and conduction bands, the light is absorbed, and the electrons are promoted from the valence band to the conduction band. The promoted electrons drift towards the n-semiconductor side and the holes move towards the p-semiconductor side as shown in Figure 9.10.

This leads to voltage and current flow from the solar cell. The number of electrons and holes that move is proportional to the light intensity.

If the light is absorbed outside the depletion region, electrons and holes need to diffuse into the p-n junction to contribute to the

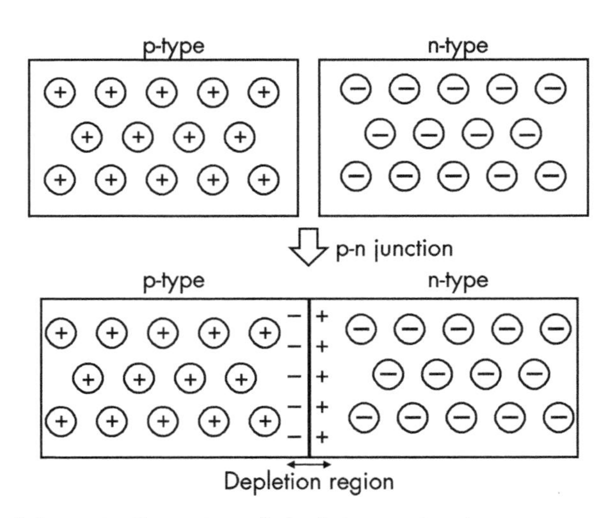


**Figure 9.9** Schematic illustration of depletion region between p-type and n-type semiconductors.

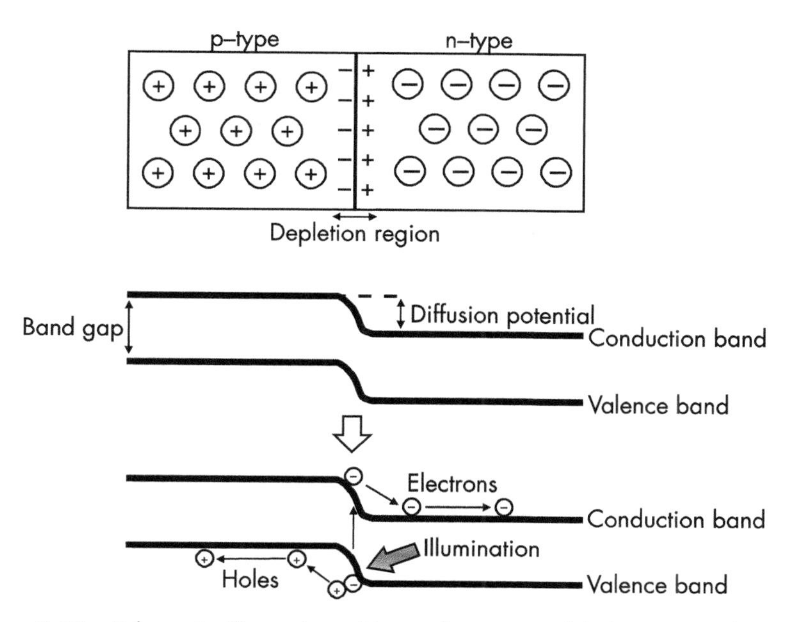


**Figure 9.10** Schematic illustration of how electrons and holes move when a p-n junction is illuminated.

current flow. However, the electrons and holes may be combined with holes and electrons, respectively, on the way to the p-n junction. This is called recombination and does not contribute to current flow. If the solar cell has impurity or lattice defects, they work as the recombination centers and decrease the solar cell efficiency. Therefore, silicon with high purity and monocrystalline structure is preferred for high efficiency.

On a side note, the excess photon energy that is greater than the band gap energy is thermalized. The solar cell temperature is also affected by ambient temperature. High temperature reduces the band gap and decreases output voltage. This decreases output power of a solar cell (e.g., –0.3% per degrees Celsius).

### 9.3.4 Structure

Figure 9.11 shows a cross-sectional image of a silicon solar cell as an example.

The base material in this figure is n-type silicon. The top of the base material is positively doped to form the p-type silicon and a p-n

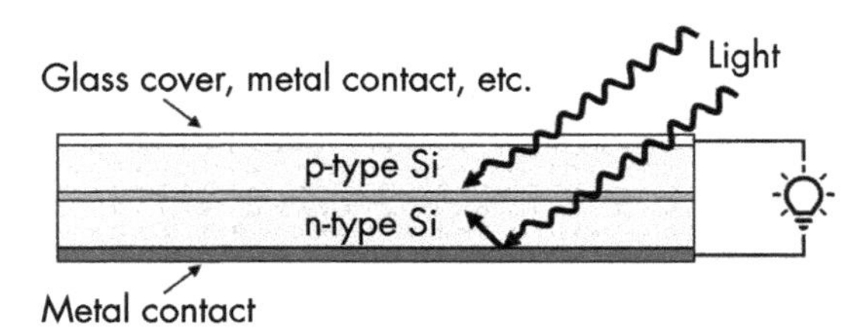


**Figure 9.11** An example of a cross-sectional image for a silicon solar cell.

junction is generated. Metal contacts are also formed on the top and bottom. As the base material of this figure is n-type silicon, this solar cell is called n-type solar cell. Note that there is a p-type silicon layer although it is called an n-type solar cell. Similarly, when the base material is p-type silicon and the top of the material is negatively doped, the solar cell is called a p-type solar cell. A solar cell product also has a highly transparent glass cover for protection, and antireflective coating to increase efficiency. In 2022, p-type silicon had a market share of ~80% because of early market penetration [6]. However, boron-doped p-type silicon, which also includes oxygen, causes a loss of performance due to boron-oxygen defects. Therefore, n-type silicon shows higher efficiency. The market share of n-type silicon is expected to increase to more than 50% by 2029 [6].

### 9.3.5 I-V Curve and Maximum Power Point

Figure 9.12 is an example of the relationship between current and voltage (I-V curve), and power output for a solar cell.

In the I-V curve, when the solar cell is at an open circuit with no current, the highest voltage is observed as OCV. The voltage decreases as the current increases and reaches zero at some point. The maximum current is called the short circuit current.

The power output, $P$, is calculated as $P = V \times I$ where $V$ is voltage, and $I$ is current. As shown in Figure 9.12, the power output of the solar cell is maximized at one point. The point is called the maximum power point (MPP). MPP depends on the light intensity and the temperature of the solar cell. There are several ICs that monitor the voltage and current of the solar cell and adjust current to optimize the output at MPP.

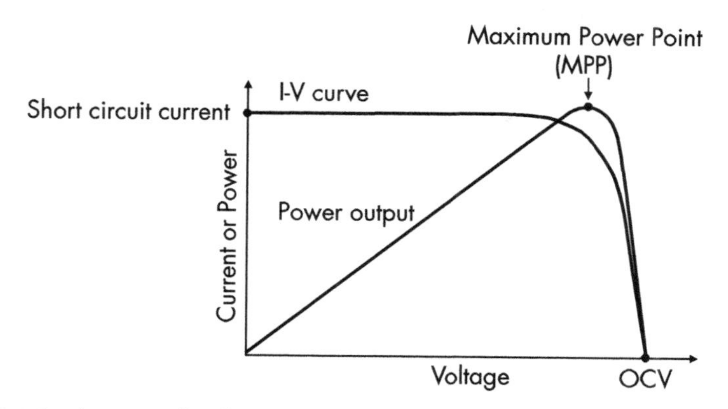


**Figure 9.12** An example of an I-V curve and power output for a solar cell

### 9.3.6 Value of Solar Cells on Electric Vehicles

Given today's market penetration, there is no doubt that solar cell technologies are successful. Still, a further assessment is needed to explore whether a system benefits from a solar cell from a practical standpoint, depending on the usage models. This chapter explains an example of a solar cell on an electric vehicle.

There is a concept to put a solar cell on the roof of an electric vehicle. When the effective area of the solar cell is $1m^2$ and the efficiency is 20%, how much energy can the solar cell on the electric vehicle roof generate when the car is left out in the sun?

For example, if the location receives 6.3 $kWh \cdot m^{-2} \cdot day^{-1}$ of solar energy on annual average [7], the possible generated energy is 6.3 $kWh \cdot m^{-2} \cdot day^{-1} \times 1m^2 \times 20\% = 1.26\ kWh \cdot day^{-1}$ on average. If the EV drives 4 miles per kWh, the solar cell extends the driving range by $1.26\ kWh \cdot day^{-1} \times 4\ miles \cdot kWh^{-1} = 5.04$ miles per day on average. This may not be attractive when the recent electric vehicles drive several hundred miles per full charge.

When the electricity cost is \$0.10/kWh, the solar cell saves 1.26 $kWh \cdot day^{-1} \times \$0.10/kWh = \$0.126$ per day. If the solar panel adds \$1,000 to the car price, it takes $\$1,000/\$0.126\ day^{-1}/365\ days \cdot year^{-1} \approx 21.7$ years to make up for the cost. This is longer than the typical product life of a car. This means that further increase in efficiency, larger panel size, and cost reduction are needed to have solar cells for electric vehicles.

On a side note, if the electric vehicle is left in the sun and the battery is heated, the battery degradation is accelerated. The battery is usually at the bottom of the car and not in direct sunlight, but it can be heated. Such degradation also needs to be considered when the solar cell is assessed for electric vehicles. There are several battery-charging algorithms that mitigate battery degradation. The details are explained in Chapter 10. These algorithms will not only reduce the battery replacement cost but also enhance the sustainability of electric vehicles.

### 9.3.7 Transparent Solar Cell

A silicon solar cell is typically opaque because it absorbs the wavelengths in the light, including the visible range, and converts them into energy as shown in Figure 9.13(a). What if a material of a solar cell absorbs only the wavelengths outside the visible range, and transmits the visible wavelengths as shown in Figure 9.13(b)? The material is transparent while working as a solar cell.

Such transparent solar cell technologies have been actively researched [8–10]. Figure 9.14 is an example of a transparent solar cell structure in a system.

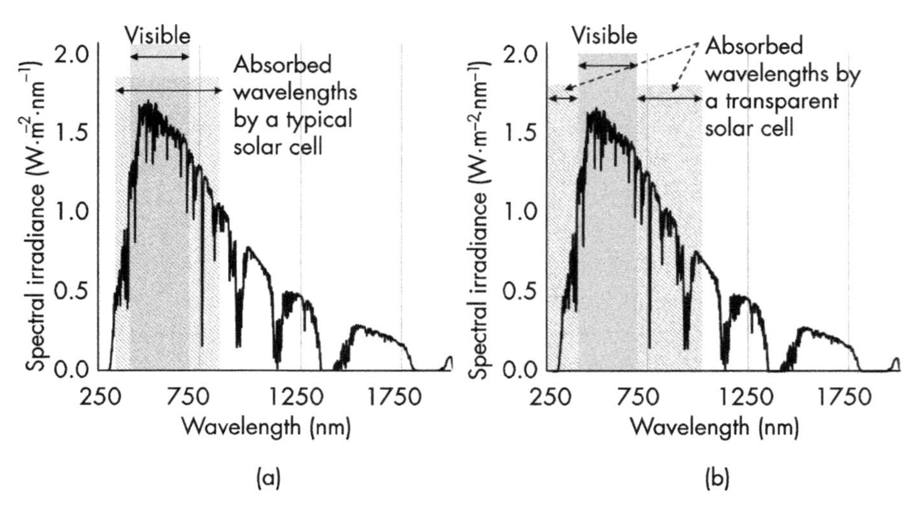


**Figure 9.13** An example of wavelengths absorbed in (a) a typical silicon solar cell (opaque), and (b) an example of a transparent solar cell. (Spectral irradiance and wavelength data is after [5].)

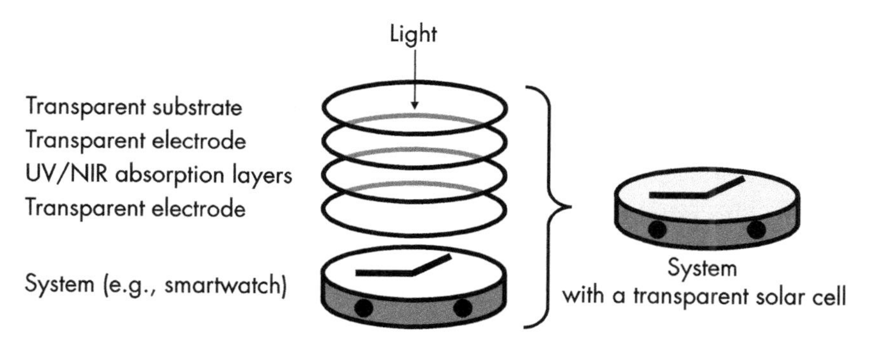


**Figure 9.14** An example of a transparent solar cell structure in a system.

In this figure, the transparent solar cell consists of a transparent substrate, transparent electrodes, and multiple ultraviolet (UV) and near-infrared (NIR) absorption layers. These are placed on a system such as a smartwatch. The transparent solar cell transmits the visible wavelengths but absorbs UV and NIR wavelengths at the junctions of the UV/NIR absorption layers (e.g., organic photovoltaic cells). The output power is supplied to the system.

This technology is expected to be used in the windows of buildings as well as in portable systems such as a smartwatch. For those usages, efficiency and transparency are the important characteristics. One example is 9.8% efficiency with 38.3% transparency [11]. In general, the transparency increases with decreasing efficiency. If high transparency is needed for the application, such as a display, low efficiency needs to be accepted. It is important to optimize the transparency and efficiency depending on the system requirements.

This book focuses on the practical applications of the technologies. Therefore, let's calculate how much the transparent solar cell contributes to battery life extension in the case of a smartwatch.

***Exercise***

You are an energy storage system engineer in a company that designs a smartwatch. You are asked how much battery life extension is expected when a transparent solar cell is installed on the watch face. Estimate battery life with the transparent solar cell based on the following conditions:

- The user is in the location that receives 6.3 $kWh \cdot m^{-2} \cdot day^{-1}$ of solar energy on an annual average. The transparent solar cell generates power only when the watch face is exposed to the sunlight. The available energy considering exposure time and light angle is 1.0% on annual average.
- The efficiency of the transparent solar cell is 5.0% in total, including all UV/NIR absorption layers.
- The area of the transparent solar cell on the watch face is 20 $cm^2$.
- There is a 1.0-Wh rechargeable battery in the watch. Battery life is 20 days without the transparent solar cell.
- System energy consumption does not change after installation of the transparent solar cell. Ignore the loss in the power delivery or battery charging. All of the generated energy by the transparent solar cell is used for the system operation.

***Answer***

1% of 6.3 $kWh \cdot m^{-2} \cdot day^{-1}$ is 0.0063 $Wh \cdot cm^{-2} \cdot day^{-1}$. With 5% efficiency and 20 $cm^2$ area, 0.0063 $Wh \cdot cm^{-2} \cdot day^{-1} \times 20\ cm^2 \times 5\% = 0.0063\ Wh \cdot day^{-1}$ is generated. The watch consumes 1.0 Wh/20 days = 0.05 $Wh \cdot day^{-1}$. The energy consumption is reduced by the generated energy from the transparent solar cell to (0.05 $Wh \cdot day^{-1}$ – 0.0063 $Wh \cdot day^{-1}$). Therefore, the battery life assumption with the transparent solar cell is 1.0 Wh/(0.05 $Wh \cdot day^{-1}$ – 0.0063 $Wh \cdot day^{-1}$) ≈ 23 days. This is a 3-day extension from the original 20 days.

* * *

The real battery life extension is smaller than this because the solar cell circuit consumes some of the generated energy. The battery life extension highly depends on the usage models. For example, if the users spend more time outdoors in the sun, they benefit more. If the display backlight needs to be brighter because the transparency of the cell is less than that of the typical cover glass, battery life extension may decrease to negative as the display backlight consumes more power. The application of the transparent solar cell to the systems needs to consider these factors.

### 9.3.8 Other Solar Cell Technologies

There are many other solar cell technologies, such as III-V solar cells, dye-sensitized solar cells, CIGS solar cells, and CdTe solar cells. III-V solar cells outperform other solar cell technologies in efficiency. III-V solar cells are opaque and consist of the elements from the group 13 (i.e., IIIA or 3A) such as B, Al, and Ga, and the group 15 (i.e., VA or 5A) such as N, P, and As. One example is gallium arsenide (GaAs). By adjusting the elements and ratio from the groups 13 and 15, the solar cell (i.e., light absorption layer) can have a different band gap that absorbs a different wavelength. When the solar cell has the multiple absorption layers with the multiple junctions, where each junction has a different band gap, the solar cell can absorb a wide range of light wavelengths including the wavelengths that a silicon solar cell cannot utilize. Also, by using concentrated light with an optical light collector, such as lenses, the efficiency can be even better. With these methods, 47.1% efficiency was achieved by National Renewable Energy Laboratory (NREL), which is the world record at the time of writing the book [12]. The III-V solar cells cost more than silicon solar cells but provide higher efficiency and better thermal/radioactive stability. Therefore, the III-V solar cells are most often used to power satellites [13].

## 9.4 ENERGY HARVESTING

Energy harvesting is the process of converting the existing ambient energy into electrical energy. For mobile systems (wearables, smartwatches, etc.) and Internet-of-Things (IoT) devices, the generated electrical energy is typically stored in a battery and contributes to extending battery life. Solar cells are part of the energy-harvesting devices and are explained in the previous section. This section covers other energy-harvesting technologies and their impact on battery life, mainly for wearables and IoT devices.

### 9.4.1 Kinetic

Energy can be generated by human movement. An electric dynamo, such as a dynamo light, is an electrical generator from kinetic energy through a magnetic field. However, the device requires coils and magnets. Therefore, it is heavy and large.

Piezoelectric device converts the mechanical stress that is driven by the kinetic energy to electrical energy via the piezoelectric effect. When the piezoelectric device is pressed and released repeatedly, it generates an AC. Such AC is converted to a DC through a rectifier and is used in a system or for battery charging. One application is to install the device in the shoes so that the device is pressed and released during walking. It was reported that the piezoelectric device where the piezoelectric material is the lead-zirconate-titanate (PZT) ceramic provided 52 $\mu$W at maximum per step with the 38 mm × 38 mm area in a shoe [14]. This is equivalent to 3.6 $\mu W/cm^2$. While the harvested power depends on testing conditions, the piezoelectric harvester typically generates the power with the same order of magnitude. It is important to understand that the generated power is not constant but at maximum. This means that even if the press and release are repeated during walking, the generated energy is not the product of the maximum power multiplied by the step count. If half power is generated on average per step, each step takes one second and several thousand steps are taken per day, the energy generated by walking is several tens of microwatt-hours per day, whereas the recent smartwatches with a bright display consume around 1 Wh per day. This means that the piezoelectric devices do little to extend the battery life of the smartwatches. The devices are suitable for low-power systems, such as sensors.

### 9.4.2 Thermoelectric Generator

A thermoelectric generator (TEG) is the device that converts heat flow to electrical energy. In a TEG, the Seebeck effect is utilized where a temperature difference between two different metals, p-type and n-type, produces a voltage difference. Figure 9.15 is an example of a structure inside a thermoelectric generator.

When a temperature difference is generated between the hot and cold sides, current flows from the n-type metal to the p-type metal through the metal contacts. This means that, to harvest energy with a TEG device, both hot and cold sides are always required. The generated energy increases as temperature difference increases. The technology is a clean and sustainable generator and has been actively researched to utilize heat waste from factories. There has been discussion on whether the heat from a laptop PC, such as a processor,

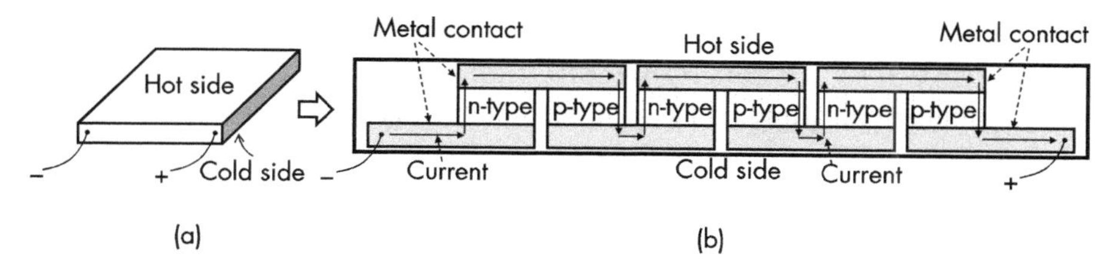


**Figure 9.15** An example of a thermoelectric generator: (a) TEG overview, and (b) an example of TEG cross-section.

can be utilized by having a TEG between the processor and the heat sink. The thermal conductivity of a TEG between a hot side and a cold side is typically low because the energy generation is efficient when the hot side is kept hot. On the other hand, the heat from the processor needs to be quickly removed to keep good performance. Placing the TEG between the processor and the heat sink makes the performance of the processor worse. There has been research to harvest energy from body temperature. For example, it was reported that a TEG, where the hot side is the body at body temperature, and the cold side is the air at 25°C, produced 83 nW/cm$^2$ (i.e., 12.5 $\mu$W) [15]. In this case, the user who wore the device was walking, and the body temperature was relatively higher than usual. When the user is not active, the generated power is less because of the smaller temperature difference between the body and the air. Also, the generated power depends on the air temperature, meaning that higher power is possible when the air temperature is low. If the user wears the device all day and the generated energy is 100 to 200 $\mu$Wh per day, what does this mean? When the recent smartwatches with a bright display consume around 1 Wh per day, the energy generated by TEG devices do little to extend the battery life of the smartwatches. However, the generated energy may be enough for low-power systems, such as sensors and low-power watches.

### 9.4.3 Radio Frequency

Radio frequency (RF) energy harvesting is a conversion of the energy from the electromagnetic field into electrical energy. The electromagnetic field in the surrounding environment, or from the transmitting coil/antenna, passes through the receiving coil/antenna in the RF energy-harvesting device. This induces an alternating electromotive force in the receiving coil/antenna by Faraday's law of induction, generating an AC in the receiving coil/antenna. Such AC is converted to DC through a rectifier, and supplied to a battery and other components in the system. The theory is the same with wireless charging, which is explained in Chapter 4. In the case of wireless charging, the electromagnetic field comes from a transmitting coil in a wireless charging pad. There has been research on energy harvesting from the existing wireless signals in the surrounding environment, such as Wi-Fi and wireless data communication for mobile phones. For efficient harvesting, the antenna design is important. For example, it was reported that the 46.7 mm × 71.7 mm antenna harvested 2.91 mW from the 2.4 to 2.5 GHz band in the ideal setup [16]. The strength of the wireless signal depends on the environment and may be weaker, resulting in less harvested energy. Similar to kinetic energy harvesting or TEG, the RF energy harvesting from the 2.4 to 2.5 GHz band does little to extend the battery life of the smartwatches with a bright display. Still, the device is enough for some low-power systems, such as sensors.

This section explained three types of energy-harvesting devices and how much power or energy is roughly expected from each type. It is important to choose the appropriate harvesting method, considering the system power/energy requirement, the device space, cost, and the usage environment. If the system is low power, an energy-harvesting device may largely extend battery life. If the device does not contribute much to battery life extension, using a larger and higher capacity battery may extend battery life longer, instead of using the space for the energy harvester.

## 9.5 HEAT TRANSFER

Battery thermal management is important because battery temperature affects battery performance, longevity, and safety. As explained in Chapter 6, over-temperature accelerates the degradation reactions in the battery, and under-temperature brings the risks of dendrite and internal short circuit during charge. While a protection IC in a battery pack can detect over-/under-temperature situations through a thermistor in a battery pack and regulates the current and voltage, it is important to avoid such situations through the system design. For example, while fast charging increases battery temperature due to joule heat caused by battery impedance, the battery temperature needs to stay within the safety limit. Such thermal design is typically the role of thermal engineers, not battery engineers. Still, if the battery engineers are familiar with the basics of heat transfer, the discussion between the battery and thermal engineers will be efficient. Therefore, this section explains the essentials of heat transfer for battery control and design.

### 9.5.1 Heat Transfer Mechanism

Heat goes from a high energy state to a low energy state. There are three modes for heat transfer: conduction, convection, and radiation.

Conduction is the energy transfer from the substance that has higher thermal energy (i.e., higher temperature) to the adjacent substance that has less thermal energy (i.e., lower temperature) through direct contact.

Convection is the energy transfer by gas or liquid flow.

Radiation is the energy emission from the substance in the form of electromagnetic waves.

The following sections explain the laws of heat transfer.

### 9.5.2 Conduction: Fourier's Law of Heat Conduction

Figure 9.16 shows a substance where the temperature on one side is $T_1$ and the temperature on the other side is $T_2$.

In this case, the rate of heat conduction can be expressed as:

$$Q_{cond} = -kA\ dT/dx \tag{9.7}$$

where $Q_{cond}$ is the rate of heat conduction in watt, $k$ is the thermal conductivity in $W \cdot m^{-1} \cdot K^{-1}$, $A$ is the surface area in $m^2$, $dT$ is the temperature change from $T_1$ to $T_2$ in Kelvin or degree Celsius, and $dx$ is the distance in meters.

This is called Fourier's law of heat conduction. This law means that the rate of heat conduction is proportional to the temperature gradient.

### 9.5.3 Convection: Newton's Law of Cooling

Figure 9.17 shows the substance where the air flows on the surface and the heat from the substance is transferred to the air.

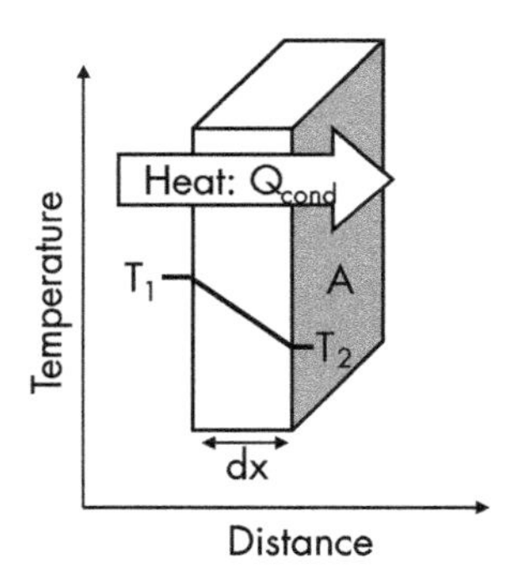


**Figure 9.16** An example of a substance with the temperature gradient.

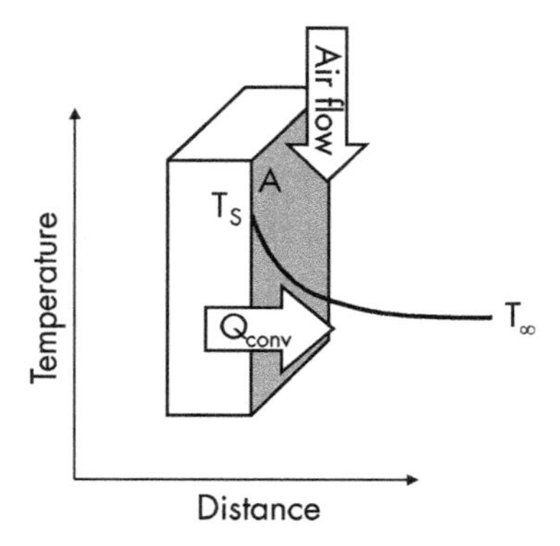


**Figure 9.17** An example of heat transfer from a substance to air flow.

In this case, the rate of heat convection is expressed as:

$$Q_{conv} = hA(T_s - T_\infty) \tag{9.8}$$

where $Q_{conv}$ is the rate of heat convection in watt, $h$ is the convection heat transfer coefficient in $W \cdot m^{-2} \cdot K^{-1}$, $A$ is the surface area in $m^2$, $T_s$ is the surface temperature in Kelvin, and $T_\infty$ is the temperature sufficiently far from the surface in Kelvin.

This is called Newton's law of cooling. This means that the rate of heat transfer is proportional to the difference in the temperatures between the substance and its surroundings.

### 9.5.4 Radiation: Stefan-Boltzmann Law

Figure 9.18 shows a substance where the surface temperature is $T_s$ and the radiation is emitted from the surface.

In this case, the heat transfer rate by radiation can be expressed as:

$$Q_{emit} = \varepsilon\sigma A T_s^4 \tag{9.9}$$

where $Q_{emit}$ is the heat transfer rate of radiation in watt, $\varepsilon$ is the emissivity of the surface where $\varepsilon$ takes the value between 0 and 1, $\sigma$ is the Stefan-Boltzmann constant, which is ~$5.670 \times 10^{-8}$ W · m$^{-2}$ · K$^{-4}$, $A$ is the surface area in m$^2$, and $T_s$ is the surface temperature in Kelvin.

This equation shows that the heat transfer by radiation is proportional to temperature to the power of four. When emissivity, $\varepsilon$, is 1, the material is called a blackbody where all of the radiation is emitted from the surface.

### 9.5.5 Thermal Modeling and Control

System components and a battery generate heat during the system operation. For example, the processors and voltage regulators in the system run with the energy from the battery and generate heat. Although it is recommended to keep the battery away from the heat sources, it may be difficult, especially for a small form-factor system that contains the components and a battery in limited space. In such a case, the heat may affect battery temperature. A Li-ion battery itself generates joule heat during charge and discharge due to the internal impedance. In addition, chemical reactions during charge and discharge also affect battery temperature. For example, for a Li-ion battery with

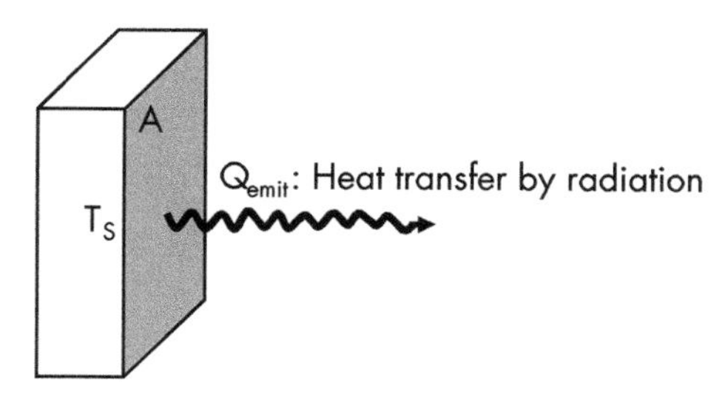


**Figure 9.18** An example of heat transfer by radiation.

$LiCoO_2$ cathode and graphite anode, endothermic and exothermic reactions are observed during charge and discharge, respectively [17].

A system and a battery pack typically communicate with each other and regulate current and voltage when over-temperature is detected. Still, it is important to consider all heat sources and avoid the over-temperature situation during normal operation by the system design.

There are many models for thermal simulations, such as one-dimensional, two-dimensional, and three-dimensional models [18]. These are based on the heat transfer modes explained earlier in this section. Some models use detailed thermal parameters, such as thermal conductivities of cathode/anode materials, separators, current collectors, and electrolytes. Large systems, such as electric vehicles, may have active or passive cooling systems such as heat-pipe and air/liquid cooling for the batteries. This is to ensure that the vehicle does not stop on the road even when the ambient temperature is high.

Figure 9.19 is a simple example of a battery thermal model with a portable battery charger.

The portable battery charger is also known as a power bank. In this example, the capacity is 6.0 Ah. For convenience's sake, it consists of a battery cell and an enclosure only. The power bank is capable of up to 2.0C fast charging. The battery cell temperature $T_{cell}$ cannot exceed 50°C during charge due to the operating temperature spec. The surface temperature of the enclosure $T_{enclosure}$ needs to be below 48°C to avoid low temperature burn.

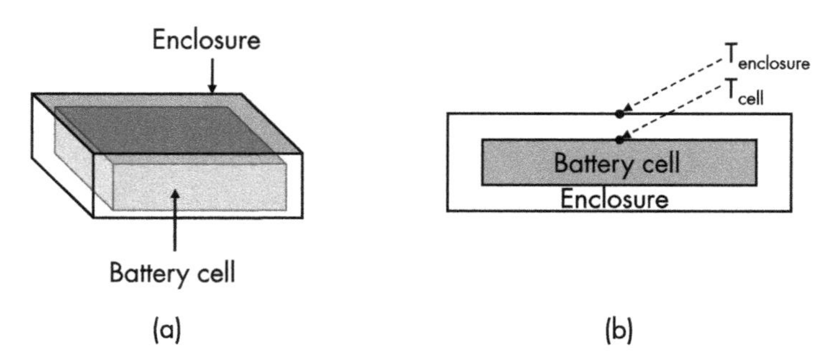


**Figure 9.19** A simple example of a battery thermal model with a portable battery charger: (a) overview, and (b) cross-section.

When cell temperature reaches 50°C, what is the possible charging current during CC charging based on the following conditions?

- The surface of the enclosure is at 48°C.
- Battery impedance is 30 mohm. During charge, the battery generates only joule heat. Ignore the heat by chemical reactions or from other heat sources.
- Thermal conductivity of the enclosure is 0.20 $W \cdot m^{-1} \cdot K^{-1}$.
- Thickness of the enclosure is 2.0 mm.
- As this is a simple example, consider only heat transfer by conduction from the top and bottom of the cell to the enclosure. The surface area for the heat transfer is 0.012 $m^2$. This includes both the top and bottom of the cell.
- The heat from the cell spreads equally across the top and the bottom of the cell.

By Fourier's law of heat conduction (9.7), the heat transfer is calculated as:

$$\begin{aligned}(\text{Heat Transfer}) &= -kA\, dT/dx \\ &= -0.2W \cdot m^{-1} \cdot K^{-1} \times 0.012m^2 \\ &\quad \times (48°\text{C} - 50°\text{C})/0.002\text{m} \\ &= 2.4\ \text{W}\end{aligned}$$

This is the heat that can be transferred from the cell. The joule heat during charge is $I^2R$ where $I$ is current, and $R$ is the impedance. As $I^2R$ needs to be lower than 2.4W to prevent the cell temperature from exceeding 50°C, the following equation needs to be met:

$$I^2 \times 0.03\ \text{ohm} \leq 2.4\text{W}$$

Solving the equation gives $I \leq \sim 8.9$A. This means that the charging current should not exceed 8.9A on these conditions although the 6.0-Ah cell is capable of 2.0C charging (i.e., 12A).

As the scope of this section is the essentials of heat transfer, this example is the simplified first step based on the specific conditions where the temperatures of both sides are defined, and only heat transfer by conduction is considered. In the real system, there may be air flow or a fluid on the surface, which requires the consideration of the convection heat transfer. Impedance that causes the joule heat may not be constant as it depends on the battery SOC and temperature. Heat transfer may also be affected by the heat of the chemical reactions and the ICs in the system. Computer-aided simulations are typically used to consider such complex conditions.

Joule heat from a battery can be mitigated not only by regulating current but also by modifying a cell design and reducing battery impedance. For example, thicker current collectors/tabs and more conductive materials in the electrodes can reduce cell impedance. Also, making the active material layers and separators thinner reduces the distance that lithium ions need to travel, resulting in impedance reduction. However, such modifications will also reduce the space for cathode and anode materials, which decreases battery capacity. It is important to understand the trade-off between impedance and the battery capacity and choose the appropriate point to meet the needs of users. For example, a small portable battery charger (e.g., 10 Wh) may be used when a smartphone battery (15 Wh) runs out. The user may need a high C-rate from the portable battery charger to meet the maximum charging current of the smartphone. In this case, the impedance reduction may be prioritized in the design of the portable battery charger at the expense of capacity reduction. If the portable battery charger is large (e.g., 30 Wh) and normal C-rate from the charger is enough for the maximum charging current of the smartphone, battery capacity may be prioritized rather than impedance reduction.

Battery charging algorithms also play an important role in mitigating battery degradation. When both battery SOC and temperature are high, battery degradation is accelerated. Chapter 6 explained the function of the battery control ICs that lowers the charging voltage to avoid high SOC when high temperature is detected. In addition to that, it is possible to predict future situations (e.g., temperature) and avoid high SOC by charging algorithms. For example, IoT devices that operate outdoors may be exposed to high ambient temperature in summer. To extend battery longevity, charging voltage and SOC can be lowered in advance by the software algorithm [19]. Such algorithms are called adaptive charging. The details are explained in Chapter 10.

## 9.6 SUMMARY

In this chapter, we learned the following essentials of battery-related technologies:

- A supercapacitor theory and energy-density comparison with a Li-ion battery.
- Solar cell technologies and their contribution to battery-life extension.
- Energy-harvesting technologies and examples of output power.
- Battery thermal control to avoid over-temperature situations.

## 9.7 PROBLEMS

### Problem 9.1

1. When a wireless device needs 100 mW for 2 seconds to transmit the data, how much energy is needed at least in joule.

2. When the required energy above is supplied by a supercapacitor where the charging voltage is 5.5V and the system accepts the voltage equal to or greater than 3.5V, how much capacitance is needed at minimum?

### Answer 9.1

1. 100 mW × 2 seconds = 100 mJ/s × 2 s = 200 mJ.

2. In this case, the capacitor energy below 3.5V cannot be used. When the capacitance is $C$, the following equation needs to be met:

$$\frac{1}{2} \times C \times (5.5\text{V})^2 - \frac{1}{2} \times C \times (3.5\text{V})^2 \geq 200 \text{ mJ}$$

Solving this gives $C \geq \sim 0.022$F. This means that ~0.022F is needed at minimum.

In reality, the loss by IR drop happens during discharge. Therefore, a higher capacitance is needed to compensate for the loss.

### Problem 9.2

1. There is an electric vehicle that has a solar cell on the roof. The effective area of the solar cell is 0.6 m$^2$ and the efficiency is 20%. How much energy can the solar cell generate when the car is left out

in the sun? The car location receives 6.3 kWh · $m^{-2}$ · $day^{-1}$ of solar energy on an annual average.

2. If the electric vehicle drives 4 miles per kWh, how many miles can the solar cell extend per day? Assume that all generated energy can be used for driving.

3. What is a potential concern to leave the electric vehicle out in the sun?

Answer 9.2

1. The generated energy is 6.3 kWh · $m^{-2}$ · $day^{-1}$ × 0.6 $m^2$ × 20% ≈ 0.76 kWh · $day^{-1}$ on average.

2. If the EV drives 4 miles per kWh, the solar cell extends the driving range by 0.76 kWh · $day^{-1}$ × 4 miles · $kWh^{-1}$ ≈ 3.0 miles per day on average.

3. (Example) If the battery is heated in the sun, battery degradation is accelerated. This decreases battery longevity and driving range on a full charge.

Problem 9.3

You are designing a system with a 9.0-Wh battery. The system operates for 30 days after full charging. The cell dimensions of the battery cell are 4.0 mm × 40 mm × 80 mm. There is an energy-harvesting device that is 1.0 mm × 40 mm × 80 mm. If the harvesting device is also integrated in the system chassis and is placed next to the battery, it generates 30 mWh per day. Determine if the energy-harvesting device should be used.

Answer 9.3

When the system operates for 30 days with a 9.0-Wh battery, it consumes 9.0 Wh/30 days = 300 mWh per day. If the energy-harvesting device is integrated and 30 mWh is generated per day, daily energy consumption is 300 mWh – 30 mWh = 270 mWh · $day^{-1}$. With this energy consumption, battery life is 9.0 Wh/270 mWh · $day^{-1}$ × 1000 ≈ 33.3 days. This is equivalent to 11% extension via (33.3 days – 30 days)/30 days × 100. However, it requires another 1.0-mm thickness on the top of the battery cell. It also requires the space and cost for the harvesting circuit. When the thickness of the 4.0-mm-thick battery cell is increased by 1.0

mm instead of using the harvesting device, the thicker battery can extend the battery life by more than 11%. Therefore, it is better not to have the harvesting device but to make the battery thicker from a battery life standpoint. Note that the answer may be different from a sustainability standpoint.

## References


[1] Pal, B., et al., "Electrolyte Selection for Supercapacitive Devices: A Critical Review," *Nanoscale Advances,* Vol. 1, 2019, pp. 3807–3835.

[2] Jabeen, N., "Enhanced Pseudocapacitive Performance of $\alpha$-MnO2 by Cation Preinsertion," *ACS Applied Materials & Interfaces,* Vol. 8, No. 49, 2016, pp. 33732–33740.

[3] Kouchachvili, L., et al., "Hybrid Battery/Supercapacitor Energy Storage System for the Electric Vehicles," *Journal of Power Sources,* Vol. 374, 2018, pp. 237–248.

[4] U.S. Department of Energy (DOE)/NREL/ALLIANCE, "2000 ASTM Standard Extraterrestrial Spectrum Reference E-490-00," https://www.nrel.gov/grid/solar-resource/spectra-astm-e490.html.

[5] U.S. Department of Energy (DOE)/NREL/ALLIANCE, "Reference Air Mass 1.5 Spectra," https://www.nrel.gov/grid/solar-resource/spectra-am1.5.html.

[6] VDMA, "International Technology Roadmap for Photovoltaic (ITRPV) 2021 Results," 13th Edition, 2022, p. 11.

[7] U.S. Department of Energy (DOE)/NREL/ALLIANCE, "Solar Resource Maps and Data," https://www.nrel.gov/gis/solar-resource-maps.html.

[8] Lunt, R., et al., "Transparent, Near-Infrared Organic Photovoltaic Solar Cells for Window and Energy-Scavenging Applications," *Applied Physics Letters,* Vol. 98, 2011, p. 113305.

[9] Stauffer, N., "Transparent Solar Cells," June 2013, MIT Energy Initiative, https://energy.mit.edu/news/transparent-solar-cells/.

[10] Traverse, C., et al., "Emergence of Highly Transparent Photovoltaics for Distributed Applications," *Nature Energy,* Vol. 2, No. 11, 2017, pp. 849–860.

[11] "Ubiquitous Energy Certifies New World Record Performance for Transparent Solar Cell," *Business Wire,* March 2019, https://www.businesswire.com/news/home/20190320005019/en/Ubiquitous-Energy-Certifies-New-World-Record-Performance.

[12] Geisz, J., et al., "Six-Junction III-V Solar Cells with 47.1% Conversion Efficiency Under 143 Suns Concentration," *Nature Energy,* Vol. 5, No. 4, 2020, pp. 326–335.

[13] U.S. Department of Energy (DOE)/NREL/ALLIANCE, "News Release: NREL Six-Junction Solar Cell Sets Two World Records for Efficiency," https://www.nrel.gov/news/press/2020/nrel-six-junction-solar-cell-sets-two-world-records-for-efficiency.html.

[14] Jeong, S. Y., et al., "Wearable Shoe-Mounted Piezoelectric Energy Harvester for a Self-Powered Wireless Communication System," *Energies*, Vol. 15, 2022, p. 237.

[15] Ren, W., et al., "High-Performance Wearable Thermoelectric Generator with Self-Healing, Recycling, and Lego-like Reconfiguring Capabilities," *Science Advances,* Vol. 7, 2021, p. eabe0586.

[16] Kumar, A., et al., "High-Isolated WiFi-2.4 GHz/LTE MIMO Antenna for RF-Energy Harvesting Applications," *AEU–International Journal of Electronics and Communications,* Vol. 141, 2021, p. 153964.

[17] Al-Hallaj, S., et al., "Thermal Modeling of Secondary Lithium Batteries for Electric Vehicle/Hybrid Electric Vehicle Applications," *Journal of Power Sources,* Vol. 110, No. 2, 2002, pp. 341–348.

[18] Lai, Y., et al., "Insight into Heat Generation of Lithium-ion Batteries Based on the Electrochemical-Thermal Model at High Discharge Rates," *International Journal of Hydrogen Energy,* Vol. 40, No. 38, 2015, pp. 13039–13049.

[19] Matsumura, N., "Battery Cycle Life Extension by Charging Algorithm to Reduce IOT Cost of Ownership," *The 34th International Battery Seminar & Exhibit,* Florida, 2017.

# 10

# BATTERY ALGORITHMS FOR LONGEVITY ESTIMATION AND EXTENSION

Long battery life with Li-ion batteries is featured in many systems, such as smartphones and laptop PCs. Users enjoy the feature on day 1 after they purchased the system. However, as time goes by, the Li-ion battery slowly but surely degrades. The battery life will not be as long in two years as it was when the battery was new. Users would like to extend the longevity if possible. However, some users may charge their smartphones overnight. Other users may always fully charge the Li-ion battery and may want to avoid low-charge state because many nickel-metal hydride batteries recommend not to fully discharge the battery. Some electric vehicle owners may charge their cars under the sun in the summer. Are these good behaviors? After reading this section, we will understand battery degradation mechanisms and have an idea of the algorithms to estimate battery longevity and extend it. Then we will be able to give the right guidance to the users whose charging behaviors are not mindful of battery longevity. Note that, unless stated otherwise, this section mainly covers a Li-ion battery with $LiCoO_2$ cathode and graphite anode that is typically used in laptop PCs and smartphones.

## 10.1 BATTERY CYCLE LIFE AND SHELF LIFE

### 10.1.1 Battery Longevity Spec

Before learning the algorithms, it is important to fully understand two battery longevity specs: cycle-life spec and shelf-life spec. These are explained in Chapter 1. Just in case, the following is a review.

An example of cycle-life spec is "At least 80% recoverable capacity after 500 cycles at 25°C." This means that battery capacity degrades through full charging and full discharging cycles at 25°C and still the battery is capable of at least 80% of initial capacity after 500 cycles. In other words, up to 20% capacity is permitted to be permanently lost.

An example of shelf-life spec is "At least 80% recoverable capacity after 12 months storage at 50% SOC at 23°C." This means that when a battery is half charged (i.e., 50% SOC) and stored at 23°C for 12 months, the battery is still capable of providing at least 80% of initial capacity with full charging.

### 10.1.2 Battery Degradation Mechanism

Why does battery capacity fade permanently? There are several reasons, which are categorized into chemical degradation and mechanical degradation in this book.

#### *10.1.2.1 Chemical Degradation*

When an electrolyte is injected in a Li-ion battery cell, it reacts with cathode and anode particles on their surfaces. This forms a resistive layer, a by-product containing lithium, as shown in Figure 10.1. The layer is also explained as SEI in Chapters 5 and 6.

The lithium in the cell is supposed to be used as part of battery capacity during charge and discharge. However, when the lithium is trapped in the resistive layer, it does not contribute to charging or discharging anymore. This results in capacity fade that is unrecoverable capacity. The resistive layer grows over time, meaning that the capacity fade increases over time.

Also, the resistive layer increases impedance of the battery. This leads to higher IR drop during discharge and battery voltage hits the discharge cutoff voltage or the system shutdown voltage earlier than

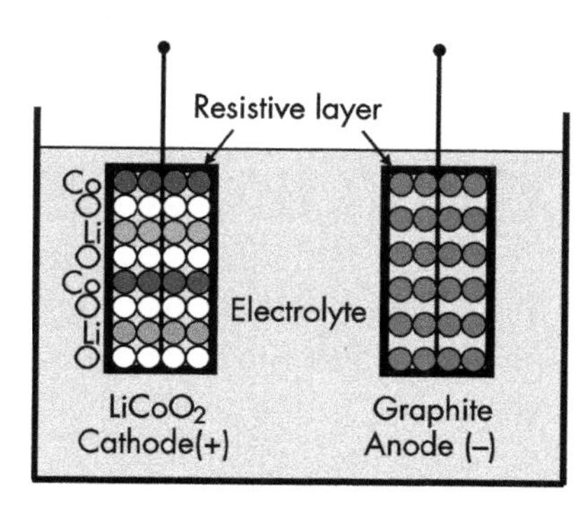


**Figure 10.1** Schematic illustration of a resistive layer on the cathode and anode.

a fresh battery. This decreases usable capacity. If discharging current is reduced, IR drop can be mitigated. However, that creates a negative impact on the system performance.

Furthermore, electrolyte decomposition happens due to unwanted chemical reactions, which generate gas. The gas prevents the lithium ions in the cell from moving between cathode and anode, causing capacity fade and impedance increase. Also, if there is not enough electrolyte after the side reactions, the cathode/anode materials not soaked with the electrolyte do not transport lithium, resulting in capacity fade.

*10.1.2.2 Mechanical Degradation*

Cathode/anode materials swell during charge and shrink during discharge due to the crystal structure change. Such volume change can cause part of the materials to delaminate from the electrodes. Those materials do not contribute to charging or discharging anymore.

The volume change also causes cracking of the resistive layer. This leads to chemical degradation. Once the cracking happens, a fresh cathode or anode surface appears and reacts with the electrolyte again. Such an unwanted reaction creates another resistive layer, consumes more precious lithium, and causes further capacity fade.

The volume change may also lead to electrode warpage. When the warpage creates a gap between the cathode and the anode and there is little electrolyte in the gap, the area does not contribute to transportation of lithium ions during charge and discharge.

#### *10.1.2.3 Acceleration in Battery Degradation*

In general, chemical degradation of Li-ion batteries is accelerated under high voltage, which is high SOC [1]. The chemical degradation is also accelerated at elevated battery temperature [1]. For example, when a smartphone is left on a table under the sun in the summer, or a battery in a laptop PC is too close to heat sources such as ICs, the battery temperature increases, and the degradation is accelerated.

Figure 10.2 shows how the temperature difference affects the cycle life.

In this chart, a 1.9-Ah cell was cycled at 25°C and another 1.9-Ah cell was cycled at 60°C. Although the cell at 60°C initially showed slightly higher capacity because of impedance reduction at high temperature, the battery degraded much faster than the cell at 25°C.

Joule heat in a battery during fast charging or high-current discharging also increases the battery temperature and causes more battery degradation.

Also, when charging/discharging cycles progress faster because of fast charging/discharging, more mechanical degradation happens.

### 10.1.3 Degradation Difference by Battery Voltages

The battery degradation rate depends on the battery voltage ranges [2–5]. Figure 10.3 shows an example of how the degradation rate changes

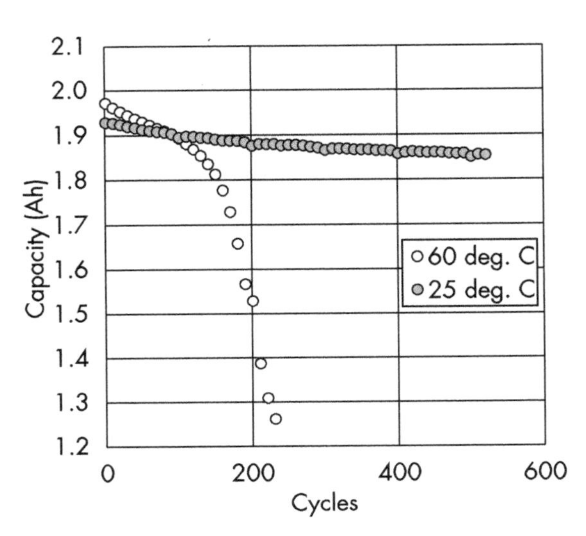


**Figure 10.2** Cycle life difference between 25°C and 60°C.

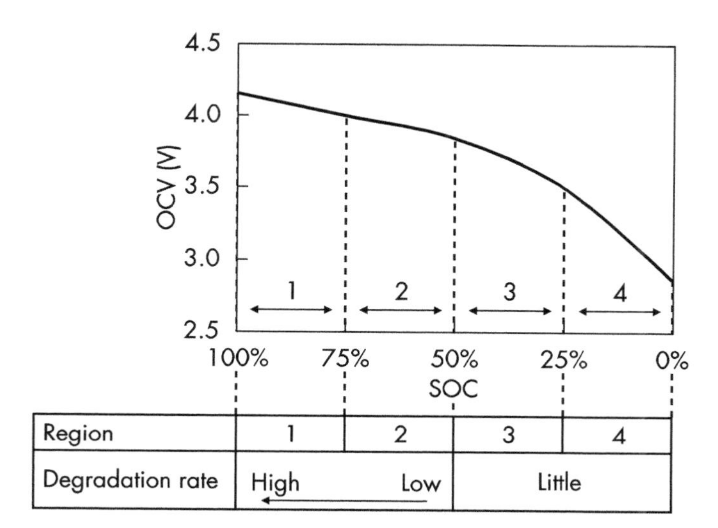


| Region | 1 | 2 | 3 | 4 |
|---|---|---|---|---|
| Degradation rate | High | Low | Little | |

**Figure 10.3** Schematic illustration of the degradation rates by charging/discharging voltages for an early Li-ion battery with $LiCoO_2$ cathode and hard carbon anode. (*After:* [2].)

by the charging/discharging voltages in the case of an early Li-ion battery with $LiCoO_2$ cathode and hard carbon anode [2]. Hard carbon anode has randomly oriented graphitic domains and is replaced with graphite later.

The curve in Figure 10.3 shows the battery OCV over SOC. The table shows the degradation rate over charging/discharging cycles in each region. For example, region 1 is the highest voltage range for the battery. When the battery is cycled in the region, the highest degradation rate is observed. Region 2 is the second highest voltage range. When the battery is cycled in the voltage range, the observed degradation rate is less than that in region 1 although the initial discharged capacity is the same with region 1. Regions 3 and 4 are below 50% SOC. In these regions, little degradation rates are observed. This indicates that the higher the battery voltage (i.e., SOC) is, the more degradation happens. This is also confirmed for a Li-ion battery with $LiCoO_2$ cathode and graphite anode that are widely used for today's smartphones and laptop PCs [3–5]. Avoiding high voltage (i.e., high SOC) extends battery-cycle life [3–5]. That also extends shelf life [5]. This is because unwanted chemical reactions are accelerated at high-voltage ranges in the case of $LiCoO_2$ cathode and graphite anode. The details of battery degradation and associated swelling are explained in the previous section and Chapter 6. By utilizing this knowledge, it is possible to enable a method to extend the battery longevity, which is explained in Section 10.3.

## 10.2 BATTERY DEGRADATION BY TEMPERATURES AND ITS ESTIMATION

### 10.2.1 Longevity Dependency on Temperature and Arrhenius Equation

The previous section explained that high temperature accelerates battery degradation. This is because unwanted chemical reactions are accelerated at high temperatures. In general, chemical reaction speed, $k$, is expressed via the Arrhenius equation below:

$$k = \mathrm{A}e^{-E/(\mathrm{R}T)} \tag{10.1}$$

where $k$ is the rate constant that is the chemical reaction speed, A is the constant, $E$ is the activation energy of the reaction, R is the universal gas constant, and $T$ is temperature in Kelvin.

### 10.2.2 Application of Arrhenius Equation to Estimate Battery Degradation

What does the Arrhenius equation practically mean? Taking the logarithm of both sides in the Arrhenius Equation gives the following:

$$\ln(k) = \ln(\mathrm{A}) - (E/\mathrm{R}) \times (1/T) \tag{10.2}$$

In this equation, ln(A) and $E$/R are constant. Therefore, this equation means that the logarithm of the degradation speed, ln($k$), is a linear function of $1/T$ that is the reciprocal temperature as shown in Figure 10.4.

In other words, it is possible to estimate the degradation speed at specific temperatures once the equation of the regression line is known. Note that the unit of temperature, $T$, is Kelvin.

### 10.2.3 Battery Degradation Estimation by Temperature

In this section, let's draw an Arrhenius plot with the sample data in Table 10.1 and estimate the degradation rate at a different temperature where data is not shown in the table.

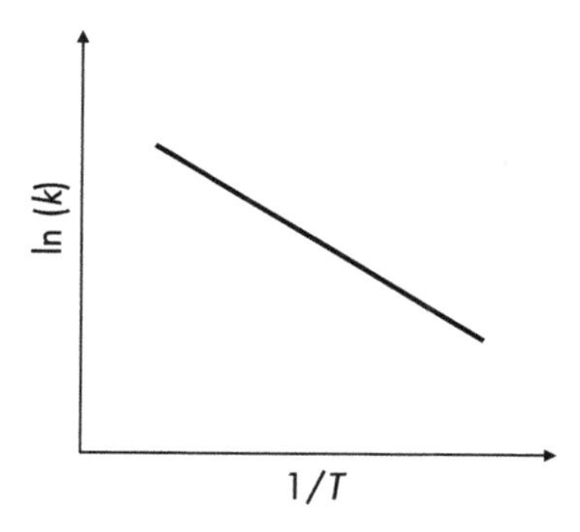


**Figure 10.4** Graphical understanding of the Arrhenius equation.

**Table 10.1**
An Example of Battery Capacity Fade in 50 Days After 100% Charge (*After:* [5])

| Temperature (°C) | Capacity Fade in 50 Days (%) |
|---|---|
| –5 | 0.32 |
| 25 | 0.97 |
| 50 | 6.04 |

Table 10.1 shows the fade of recoverable capacity after the battery samples are charged to 100% SOC and each sample is stored at the respective temperature for 50 days. First, copy the table into your spreadsheet, such as Microsoft Excel. Then follow the steps below.

*Step 1:* In the new column, calculate $T$ in Kelvin. Temperature in degrees Celsius + 273.15 equals the temperature in Kelvin.

*Step 2:* In the new column, calculate $1/T$.

*Step 3:* In the new column, calculate $\ln(k)$, the natural logarithm of $k$, where $k$ is the capacity fade in 50 days.

*Step 4:* Plot $\ln(k)$ over $1/T$.

After taking step 1 through step 3, you will obtain Table 10.2.

**Table 10.2**
An Example of Battery Capacity Fade with Necessary Columns for an Arrhenius Plot

| Temperature (°C) | Temperature (K) | 1/T ($K^{-1}$) | *k*, Capacity Fade in 50 days (%) | ln(*k*) |
|---|---|---|---|---|
| -5 | 268.15 | 0.0037 | 0.32 | –1.14 |
| 25 | 298.15 | 0.0034 | 0.97 | –0.03 |
| 50 | 323.15 | 0.0031 | 6.04 | 1.80 |

When the natural logarithm of the capacity fade in 50 days, $\ln(k)$, is plotted over $1/T$, Figure 10.5 will be obtained with the equation of the regression line.

Next, let's estimate the capacity fade at 35°C in 50 days. The temperature 35°C equals 35 + 273.15 = 308.15 K. When the temperature value is inserted in the equation of the regression line in Figure 10.5, $\ln(k)$ is calculated as ~0.87. This means that the capacity fade at 35°C in 50 days is $e^{0.87} \approx 2.4\%$.

On a side note, in general this method is used when the reactions of the chemical degradation are the same over the temperatures. When a different reaction, such as electrolyte decomposition at high

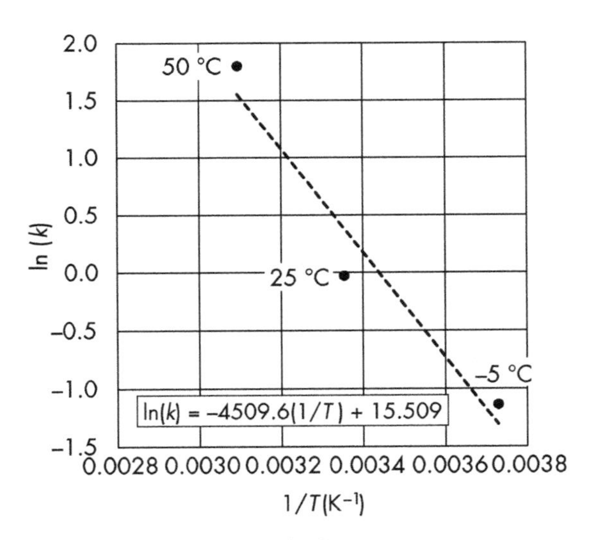


**Figure 10.5** Arrhenius plot of the sample data.

temperature happens, the different rate of the capacity fade needs to be considered.

## 10.3 LONGEVITY EXTENSION BY ADAPTIVE CHARGING

### 10.3.1 Introduction of Adaptive Charging

We already understand the battery degradation mechanisms from Section 10.1. However, that is not enough. The goal of this book is to utilize the knowledge and explain a solution to the problem. Therefore, this section explains how to extend battery longevity based on the degradation knowledge. Longevity extension is beneficial because that leads to less battery replacement, resulting in lower cost of ownership (CoO). CoO is important for any system that use batteries, such as smartphones, electric vehicles, and IoT devices. It is especially important for electric vehicles and IoT devices. Battery cost of electric vehicles is high. Nobody would doubt longevity extension lowers CoO for electric vehicles. For IoT devices, battery cost may not be high. However, such devices may be placed on the top of a mountain, in the middle of a desert, or in outer space. Sending a service personnel there to replace the battery costs a lot. Therefore, battery longevity extension helps IoT devices reduce CoO.

Chapter 1 explained that the full-charge capacity of a battery is decreased after charging and discharging are repeated. Figure 10.6 is an example of battery cycle life.

When the battery is new, where the cycle number is 1, users may enjoy as long of a battery life as the system spec shows. For example, in the case of a smartphone, if the spec claims 10 hours of talk time, users can enjoy 10 hours on day 1. However, slowly but surely, battery degradation proceeds. In Figure 10.6, battery capacity decreased to 88% after 500 cycles. This means that talk time becomes less than 9 hours after 500 cycles, which may be in ~1.4 years if one cycle is performed a day and no other degradation than cycling is considered.

In reality, such degradation is accelerated at high battery voltage (i.e., high SOC) and at high temperature. Mixing these factors degrades battery longevity even more. Then, how can we extend battery longevity such as cycle life? The answer is an adaptive charging algorithm. The next three sections explain example algorithms.

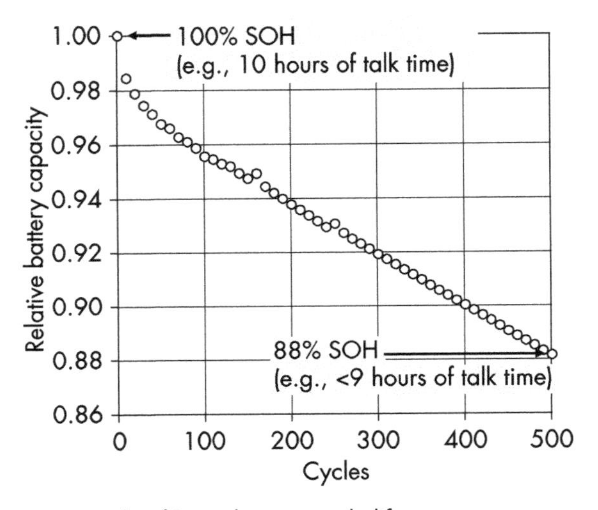


**Figure 10.6** An example of Li-ion battery cycle life.

### 10.3.2 Adaptive Charging by Scheduling Application

This section explains the adaptive charging algorithm by a scheduling application. The concept is simple. It basically charges the battery as needed by learning from a scheduling application. Figure 10.7 shows an example of battery usage with normal charging and adaptive charging.

Normally a battery is charged to 100% and is used to lower SOC (e.g., 0%) as shown in the left bar. In the next charge cycle, the battery is fully charged again and is used. Instead of doing so, if the system can accept 90% capacity until the battery is charged next time, charging the battery to 90% and using it as shown in the right bar avoids high-charge level that is a high-degradation region as explained in Section 10.1. As a result, battery longevity can be extended. Limited charging (e.g., 90% charging) can be achieved, for example, by lowering charge cutoff voltage, or by monitoring SOC information in a fuel gauge IC and stopping charging at the desired SOC.

Figure 10.8 is the cycle test results with and without adaptive charging [6].

In this case, the system needed 90% charging. Therefore, the battery with adaptive charging was charged to 91%, which is slightly higher than 90% to avoid sudden system shutdown. The other curve is without adaptive charging where the battery was always charged

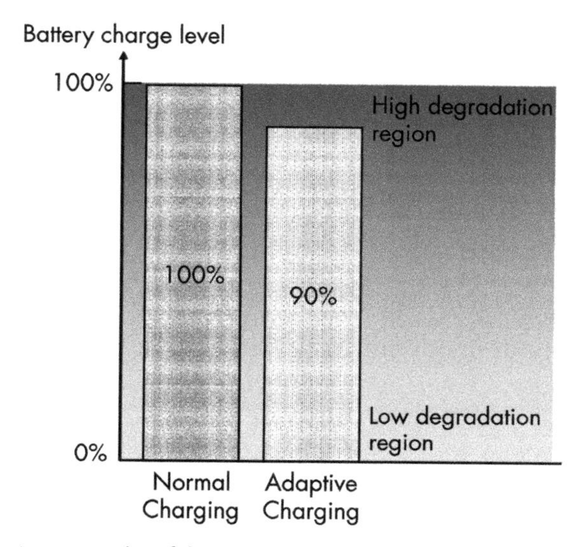


**Figure 10.7** An example of battery usage with normal charging and adaptive charging.

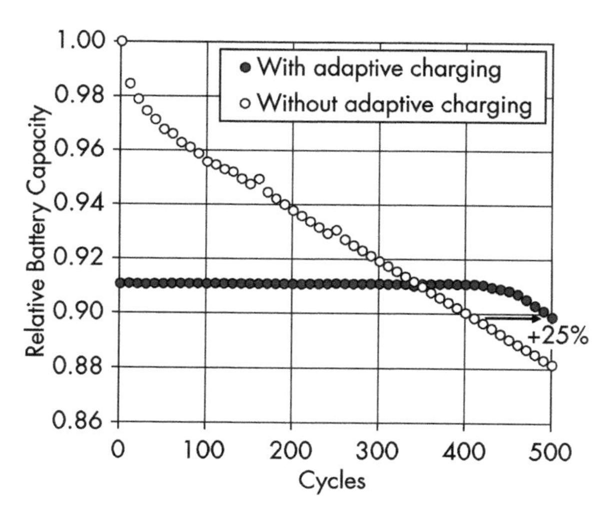


**Figure 10.8** Cycle life with and without adaptive charging.

to 100%. Initially the plots without adaptive charging showed higher capacity as the battery was charged more. However, the battery degraded faster. As a result, the adaptive charging delayed degradation and extended the cycle life by 25%.

This means that the lower the charge level is, the better cycle life (i.e., longevity) is expected in the case of batteries with $LiCoO_2$ cathode and graphite anode. The algorithm can be applied to other systems if the battery uses the same cathode and anode chemistries. For example, a cleaning robot may clean a room based on the preset schedule. If one cleaning job requires 70% capacity, charging the robot to 70% (plus some to be on the safe side) extends battery longevity than charging to 100%. Another example is an electric vehicle. It may need 80% charge to satisfy daily usage. In such a case, charging to 80% (plus some to deal with the unexpected situations such as traffic congestion) extends battery longevity than charging to 100%. This contributes to a significant reduction in CoO and enhances sustainability because of less battery replacement.

On a side note, as a battery degrades over time, continuous optimization of charging percentage is needed. Charging level optimization can be performed by monitoring SOC in a fuel gauge IC and stopping charging at the desired SOC, or adjusting charging cutoff voltage. Leveraging SOC information provides more precise and adaptive control than adjusting charge cutoff voltage.

### 10.3.3 Adaptive Charging Through Overnight Charging: Delayed Charging

Some users may charge their smartphones overnight every day. This behavior decreases battery longevity. This is because battery charging typically completes within 2.5 hours whereas users sleep much longer, often 6 to 8 hours. This means that the battery stays long at 100% SOC that is in the high-degradation region. To avoid this, some recent smartphones delay the 100% charging. For example, it starts normal charging when a charger is plugged into the smartphone and a user goes to bed. When the battery charge level reaches a certain level, such as 80%, it pauses charging. Then, it resumes charging before the user wakes up so that the user sees 100% on wakeup. This reduces time at high-charge level (i.e., high degradation region explained in Section 10.1) and extends battery longevity.

### 10.3.4 Adaptive Charging by Situations: Situational Charging

Adaptive charging can also be enabled by considering situations. For example, some IoT devices may operate with a battery charged by a solar cell. Such devices are charged by a solar cell in the daytime and are operated only by a battery at night. In winter, full battery charging may be needed because the daytime is short, and the nighttime is long. However, in summer, full charging may not be needed because the daytime is longer, and the nighttime is shorter as shown in Figure 10.9. Avoiding full charging by situations (e.g., season, usage), especially in summer when the battery of the IoT device becomes hot, can extend battery longevity [7]. IoT devices, such as smart meters that monitor environmental information with a battery and a solar cell, can benefit from this method.

## 10.4 SUMMARY

In this chapter, several adaptive charging algorithms were explained. These are the applications of battery degradation knowledge. This means that finding new degradation mechanisms leads to further invention of adaptive charging algorithms. For example, arbitrary fast

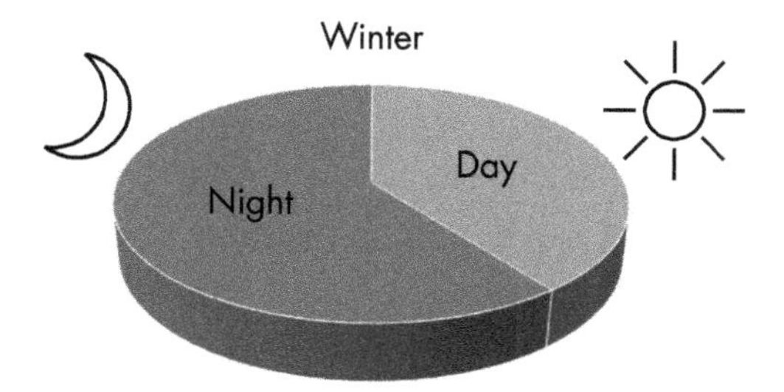


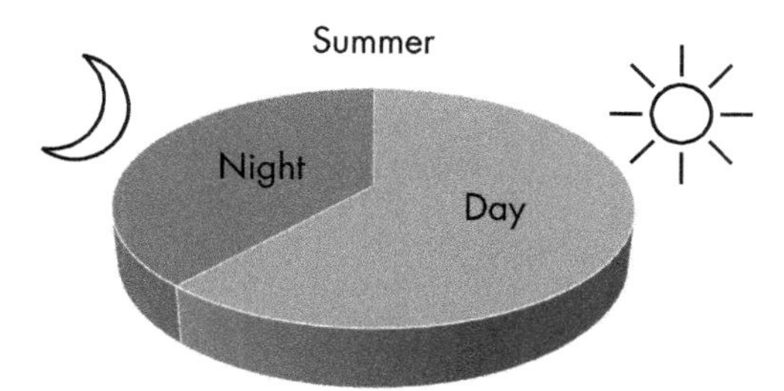


**Figure 10.9** An example of daytime and nighttime difference between winter (top) and summer (bottom).

charging accelerates battery degradation, but appropriate charging algorithm, such as step charging, explained in Chapter 4, mitigates the degradation and extends battery longevity. Application of machine learning/deep learning to battery charging can also contribute to longevity extension, which is explained in Chapter 12. It would be great if this book helps the innovation of future algorithms and bring humans a further sustainable world.

## 10.5 PROBLEMS

### Problem 10.1

Draw an Arrhenius plot with the sample data in Table 10.3 and estimate capacity fade at 35°C in 50 days.

### Answer 10.1

When the same steps in Section 10.2.3 are followed, the Arrhenius plot shown in Figure 10.10 is derived.

From the regression line in the figure, the natural logarithm of the capacity fade at 35°C in 50 days is calculated as $-5176.6 \times 1/(35 + 273.15) + 17.176 \approx 0.38$. Therefore, the answer is $e^{0.38} \approx 1.5\%$.

### Problem 10.2

There is a cleaning robot that runs with a Li-ion battery with $LiCoO_2$ cathode and graphite anode. It needs 70% charge to finish a daily cleaning job. Which of the following charging methods will result in better battery cycle life?

A: Charge the battery to 100% and use it until 30%. Repeat the charging pattern.

**Table 10.3**
Another Example of Battery Capacity Fade in 50 Days After 50% Charge (*After:* [5])

| Temperature (°C) | Capacity fade in 50 days (%) |
|---|---|
| 25 | 0.83 |
| 50 | 3.18 |

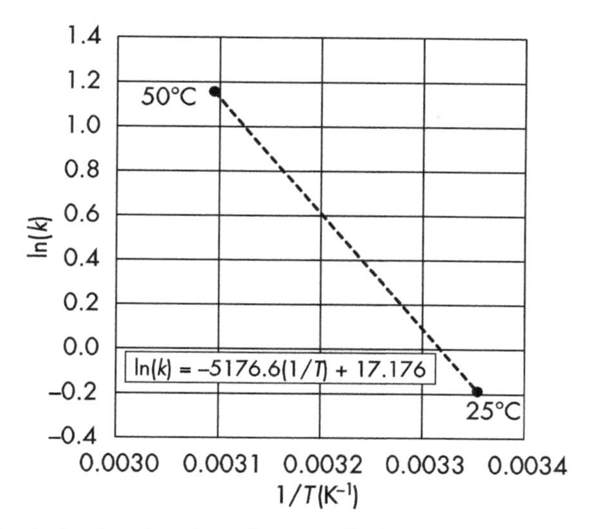


**Figure 10.10** Arrhenius plot of another sample data.

B: Charge the battery to 80% and use it until 10%. Repeat the charging pattern.

Answer 10.2

B will show better cycle life because the lower charging percentage, which is lower charge voltage, causes less degradation for a Li-ion battery with $LiCoO_2$ cathode and graphite anode.

Problem 10.3

When electric vehicles emerged, some vehicles employed the same Li-ion batteries as used in laptop PCs. The cathode is $LiCoO_2$ and the anode is graphite. The cycle-life spec of the battery is typically 500 cycles. However, electric vehicles need thousands of cycles. Explain a possible method to enable the longer cycle life with the same batteries.

Answer 10.3

(Example) While laptop PCs charge the battery to 100% during cycles, the electric vehicles may charge the battery to less than 100% (e.g., 70%) and define the threshold as their full charging. As this avoids high charging percentage, which is a high-degradation region, this extends cycle life.

## References

[1] Edge, J., et al., "Lithium Ion Battery Degradation: What You Need to Know," *Physical Chemistry Chemical Physics,* Vol. 23, 2021, pp. 8200–8221.

[2] Takei, K., et al., "Cycle Life Estimation of Lithium Secondary Battery by Extrapolation Method and Accelerated Aging Test," *Journal of Power Sources,* Vols. 97–98, 2001, pp. 697–701.

[3] Saxena, S., et al., "Cycle Life Testing and Modeling of Graphite/$LiCoO_2$ Cells Under Different State of Charge Ranges," *Journal of Power Sources,* Vol. 327, 2016, pp. 394–400.

[4] Choi, S., et al., "Factors that Affect Cycle-Life and Possible Degradation Mechanisms of a Li-ion Cell Based on $LiCoO_2$," *Journal of Power Sources,* Vol. 111, No. 1, 2002, pp. 130–136.

[5] Sun, Y., et al., "Derating Guidelines for Lithium-Ion Batteries," *Energies,* Vol. 11, No. 12, 2018.

[6] Matsumura, N., "Battery Cycle Life Extension by Charging Algorithm to Reduce IOT Cost of Ownership," *The 34th International Battery Seminar & Exhibit,* Florida, 2017.

[7] Matsumura, N., "Selection of Battery Chemistry and Charging Algorithm for IOT Devices," *The 35th International Battery Seminar & Exhibit,* Florida, 2018.

# 11

# BATTERY APPLICATION TO VARIOUS SYSTEMS

Li-ion batteries are used everywhere, in smartphones, wearables, drones, electric vehicles, and more. Each system has its unique usage. Therefore, the expectation of the batteries depends on the systems. This section explains the typical requirements for the batteries of the popular systems.

## 11.1 WEARABLES

### 11.1.1 Battery Usage in Wearables

Smartwatches are one of the popular systems in wearable devices. There are many features in smartwatches, such as step pedometer, heart rate monitor, GPS, always-on display, and built-in cellular. While these features enhance user experience, each feature consumes battery energy. A display panel and its control system take ~60% of overall system power on average [1]. Also, a built-in cellular needs sporadic high power when transmitting and receiving data. These cause large IR drops, hit the system shutdown voltage early, and result in shorter battery life. While users desire many features in the smartwatch, long battery life is still necessary. Then, does using a high energy-density battery address the challenge?

### 11.1.2 Method to Extend Battery Life

In reality, a high energy-density battery may not always provide long battery life. In general, higher energy-density batteries have higher impedance because of internal compaction of electrodes and less surface area of active materials. High impedance may also come from cell designs, such as thicker electrodes and/or thinner current collectors/terminals, which are intended to increase energy density. Figure 11.1 shows an example of the comparison for the simulated discharging curves between a high energy-density and high impedance cell (battery A) and a low energy-density and low impedance cell (battery B).

In this figure, battery A is the high energy-density cell with 400 mAh. Battery B is the same-sized simulated cell with 14% less energy density and 42% less impedance during both continuous and pulse discharges, compared to battery A. If the smartwatch discharges the cells at 0.2C and shuts down at 3.4V, battery A provides more capacity than battery B as shown at point a and b in Figure 11.1. This is because battery A has higher energy density than battery B. However, when the display and its control system require high current (e.g., continuous 1C) and the built-in cellular periodically extracts sporadic

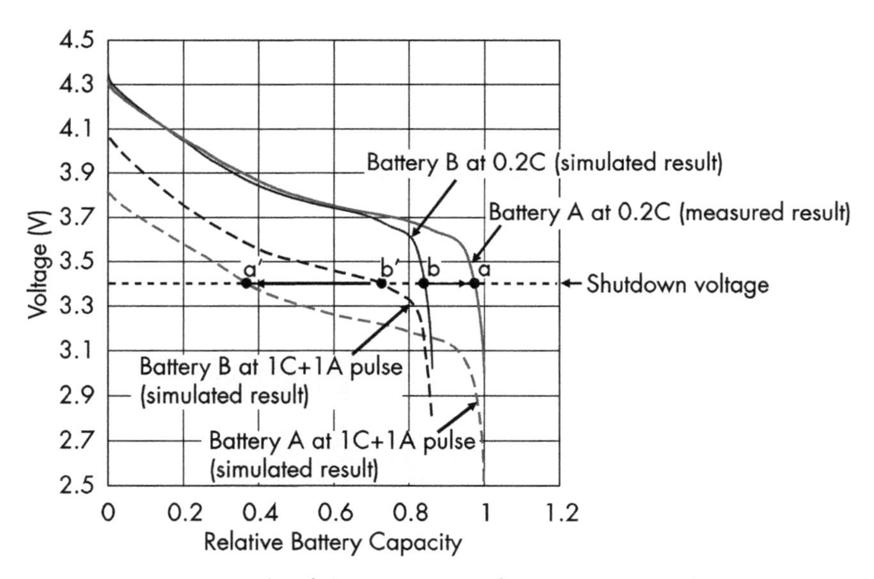


**Figure 11.1** An example of the comparison for the simulated discharging curves between a high energy-density and high impedance cell (battery A), and a low energy-density and low impedance cell (battery B).

high current (e.g., 1A pulse), the voltages of both discharging curves drop due to IR as the dashed lines show. In this case, battery A and B shut down at point a′ and b′, respectively. This means that battery A provides less capacity than battery B because battery A has higher impedance although it has higher energy density. The battery size of smartwatches is smaller than that of smartphones or laptop PCs. As the smaller battery has higher impedance because of limited electrode areas, IR drop affects battery life more in smartwatches than smartphones or laptop PCs in the same current. To enable longer battery life in a system with a small battery, such as wearables including smartwatches, it is important to balance energy density and impedance, depending on the uses that the product is expected to support, such as power requirement and its duration.

On a side note, battery energy density and impedance largely change, depending on the cell design/structure, especially when technologies evolve. For example, if a liquid electrolyte is replaced with a solid-state electrolyte to enable a next-generation lithium metal anode, high energy density can be expected but impedance may also increase as explained in Chapter 5. If electrodes have multiple connections to terminals, such as a stacking structure explained in Chapter 6, impedance will be largely reduced. This section emphasizes the importance of impedance reduction as well as new chemistry developments leading to battery life extension.

## 11.2 SMARTPHONES, TABLETS, AND LAPTOP PCS

### 11.2.1 Battery Usage in Portable Systems

Portable systems such as smartphones, tablets, and laptop PCs use high energy-density batteries because long battery life is desired. The systems also have high-performance processors which extract sporadic high current from the batteries and enhance the system response as needed. However, high discharging current from the high energy-density batteries that have high impedance causes large IR drop. Unlike the batteries in wearables, the batteries for portable systems are larger and relative impedance is smaller. Still, when battery voltage under IR drop hits system shutdown voltage, or discharge cutoff voltage, the system shuts down. Systems are typically designed to avoid early and sudden shutdown. For example, having a 2S or 3S battery pack in a laptop PC provides higher voltage, which gives larger voltage room to the system shutdown voltage. However, battery impedance increases at low temperature. It also increases when the battery is degraded. The systems, especially the small systems such as smartphones that have a 1S battery, need a function that fundamentally avoids the sudden system shutdown.

### 11.2.2 Method to Avoid Sudden System Shutdown and Extend Battery Life

One way to avoid the sudden system shutdown is to estimate battery impedance and prepare for the coming shutdown. For example, if the battery impedance is modeled with an equivalent circuit model, explained in Chapter 3, the system can estimate how high current hits the system shutdown voltage and regulate the discharging current before hitting the shutdown voltage. This reduces IR drop, delays hitting the shutdown voltage, and extends battery life. For example, some systems may switch the operation to the battery-saving mode at low SOC (e.g., 20% SOC) with or without notification to the users. This mode regulates the discharging current by dimming the display light, limiting processor performance, turning off nonurgent background activities, and so forth. While the response in the system may become slow under the mode, it extends battery life and may be beneficial to some users who need the last-minute call, email check, or document editing.

In addition, systems typically start the safe shutdown process, for example, at 3% SOC in advance of the system shutdown. This will protect the data.

## 11.3 DRONES

### 11.3.1 Battery Usage in Drones

Drones use Li-ion batteries. Their flight time is typically around 20 minutes. This is much shorter than the operation time of the portable systems with Li-ion batteries, such as 10 hours of a smartphone. This is because the motor of a drone draws high current from the battery. The 20-minute flight time means that the discharging C-rate is 3C on average. When the drone is ascending, it requires even higher current than hovering.

### 11.3.2 Requirements for Drone Batteries

It is necessary to extend the flight time. If a battery with high-gravimetric energy density is used, it reduces the weight of the battery pack to provide the same energy and seems to extend the flight time. However, the impedance of the high energy-density battery is typically high. Providing the high current for the drone flight causes a large IR drop. Lowering the battery impedance is a trade-off against the energy density. Therefore, it is important to balance the gravimetric energy density, impedance, and battery weight for the optimized usable energy that leads to longer flight time.

Also, during flight, the temperature of the drone battery increases because of joule heat associated with the battery impedance. Figure 11.2 is an example of drone battery temperature during continuous discharge at 2.5C from a battery cell [2].

In this case, battery temperature at the end of discharging was around 33°C. In general, when battery temperature increases, impedance decreases because of better ionic mobility. Still, the figure shows continuous temperature increase during discharge even after temperature increases and impedance is decreased. This is because heat generation is faster than heat dissipation from the battery. If the baseline impedance is higher, temperature will increase more. Therefore,

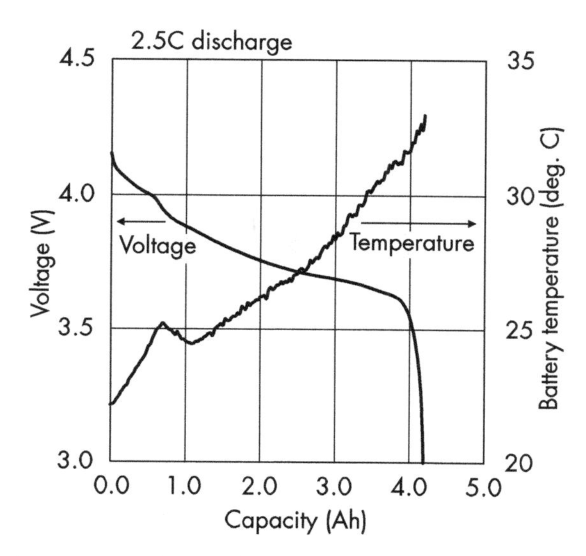


**Figure 11.2** An example of drone battery temperature during discharge at 2.5C.

it is required to make sure that the battery temperature is within the safe operation range during the flight. Also, the battery, especially the electrolyte, needs to be durable against high temperature to mitigate battery degradation.

Some drones need to fly in a cold environment. When the battery temperature is low, battery impedance increases, resulting in a larger IR drop. Once the drone flies, battery temperature may increase because of the joule heat and IR drop may decrease. This means that the IR drop, especially during takeoff, needs to be small enough to avoid hitting the system shutdown voltage.

## 11.4 IOT DEVICES

### 11.4.1 Example of IoT Batteries

IoT devices are diversified, such as tracking systems for logistics, smart thermostats, and environmental sensors. There are many things to consider in battery selection. For example, should the battery be rechargeable or nonrechargeable? What kind of battery chemistry is suitable, such as Li-ion, Ni-MH, or alkaline? How much is the cost of ownership that includes the battery cost and longevity?

### 11.4.2 Batteries for IoT Devices and Consideration in Selection

Table 11.1 is the comparison of typical batteries [3]. Note that the data in the table are representative values and may vary by manufacturers, models, sizes, temperature, and other factors.

Each battery has its own advantage. For example, rechargeable Li-ion batteries and nonrechargeable lithium coin cells provide high energy-density and high voltage. Alkaline batteries are standardized in dimensions such as AA and AAA batteries. Alkaline batteries are also affordable and available in many places near the users. Ni-MH batteries provide a similar voltage to the alkaline batteries, and thus are compatible with the alkaline batteries in many systems. The Ni-MH batteries also provide further affordability with their rechargeability, compared to the nonrechargeable alkaline batteries.

In general, lithium-based batteries such as rechargeable Li-ion batteries, and nonrechargeable lithium coin cells, are suitable for devices that require high energy and power. Non-lithium-based batteries are used for low-power devices.

**Table 11.1**
Comparison of Typical Batteries for IoT Devices

| | **Rechargeable** | | **Nonrechargeable** | |
|---|---|---|---|---|
| | **Li-ion** | **Ni-MH** | **Lithium Coin Cell** | **Alkaline** |
| **Energy Density** | Baseline (700+ Wh/l) | ~½ of Li-ion | Higher | Lower (250~500 Wh/l) |
| **Nominal Voltage** | ~3.8V | 1.2V | 3V | 1.5V |
| **Cycles** | Several hundreds to thousands | Several hundreds to thousands | n/a | n/a |
| **Self-Discharge** | 0.3%~1.2%/ month | ~3%/month | <0.1%/month | ~0.2%/month |
| **Cost (Online Volume Price)** | ~$0.2 /Wh (*18650) + protection + charger | <$1.0 /Wh (*AA) + charger | ~$0.2 /Wh (*CR2032) | <$0.1 /Wh (*AA) |

Then, which battery is suitable for the normally low-power but sometimes high-power devices?

The answer depends on the required power and battery impedance. Figure 11.3 shows how the voltage of a lithium coin cell (CR2450) changes under various discharging currents.

In the figure, the coin cell where the spec shows the 540-mAh rated capacity, was at 3.07V as OCV. It was first discharged at

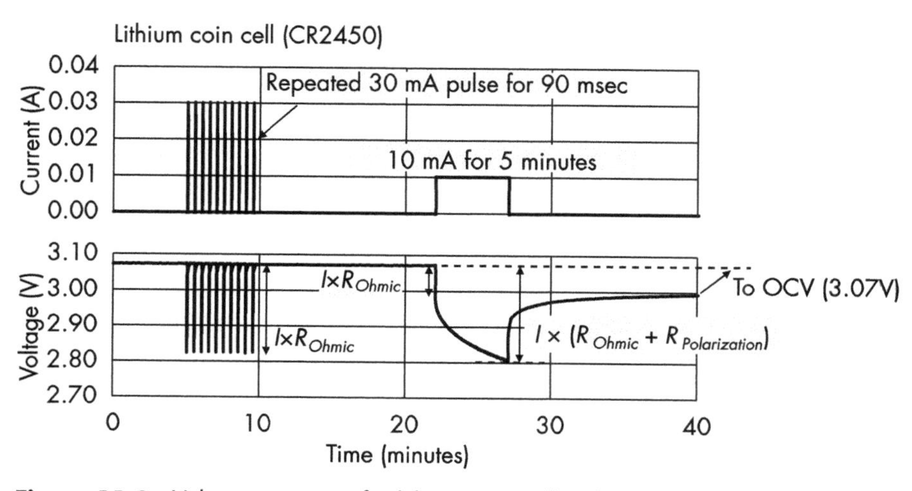


**Figure 11.3** Voltage reaction of a lithium coin cell under various currents.

short- and high-discharging currents that was a repeated 30-mA pulse for 90 msec once every 30 seconds. Then, the cell was discharged at a long- and low-discharging current that was 10 mA for 5 minutes. After that, the battery voltage under no current was slowly recovered to 3.07V, which is the same as the original OCV. As explained in Chapter 3, battery voltage drops during discharge because of IR drop. The voltage drop is expressed as $I \times R$ where $I$ is the discharging current, and $R$ is the battery impedance.

When the 30-mA pulse current was applied, the battery voltage dropped by $I \times R$. When the 10-mA continuous current was applied, the initial voltage drop was almost one-third of the voltage drop at 30 mA. This is because 10 mA is one-third of 30 mA. However, the voltage kept decreasing during the 10-mA discharge and was finally lower than the voltage at 30 mA. This is because the battery impedance increased during discharge although the discharging current is lower. The battery impedance ($R$) consists of ohmic portion ($R_{ohmic}$) and polarization portion ($R_{polarization}$), which are resistance and capacitive reactance, respectively. In a simple model, $R$ can be written as:

$$R = R_{ohmic} + R_{polarization} \tag{11.1}$$

When the discharging current is applied, the ohmic portion of the impedance reacts immediately, causing IR drop by $I \times R_{ohmic}$. During discharge, the polarization portion of impedance increases due to the capacitive reactance in the battery, and thus the IR drop by $I \times R_{polarization}$ increases. The details of impedance models are explained in Chapter 3. The system shuts down when the dropped voltage hits the system shutdown voltage. Therefore, it is important to understand the discharging profile of the system and choose the right battery with the right impedance to maximize the battery life.

Figure 11.4 is the comparison of the usable battery capacity among different CR2450 models [3].

In this figure, CR2450 coin cells from different manufacturers and models were discharged at the same discharging profile where 32 mA for 2 sec and 50 $\mu$A for 5 sec were repeated until the battery voltage hit 2.2V, as an example of a use case scenario. The chemistry and dimensions are the same among the cells and the rated capacities in the spec are similar, around 540 to 620 mAh. However, the usable capacities were largely different for the cells. For example, the model

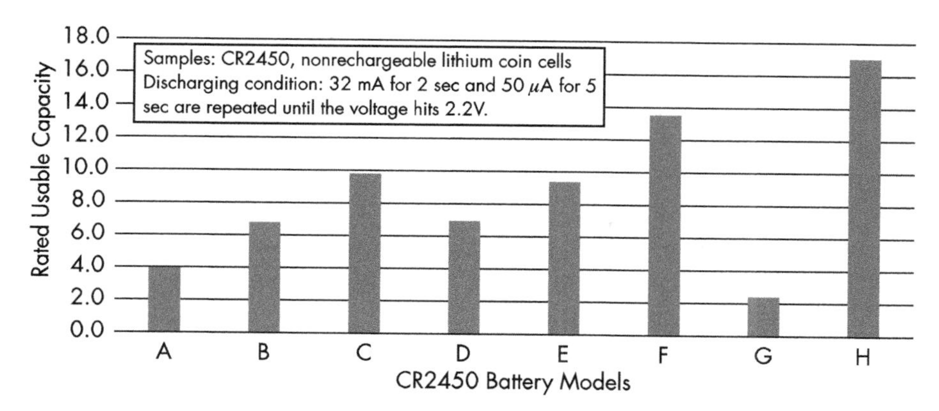


**Figure 11.4** Usable capacity comparison among different CR2450 models.

H showed 7 times more usable capacity than the model G. This is because of the difference in the impedances of the cells. Some hit 2.2V early due to higher impedance. If the discharging current is always low, the cells will show similar capacities because of the less impact by IR drop. The rated capacities of some cells are higher than others because of compaction of the electrodes in the cells. Such cells may be intended to extend battery life of low-power devices, and the price may be higher. However, these cells show higher impedance due to the internal compaction. When the discharging profile includes high current, the battery life is largely influenced by impedance as Figure 11.4 shows. For battery selection, it's important to understand the discharging profile of the device, impact of battery impedance, and cost of ownership, including the initial battery cost, longevity, and battery replacement cost.

## 11.5 BACKUP/STATIONARY BATTERY

### 11.5.1 Examples of Backup/Stationary Battery

Backup/stationary batteries are designed to support the continuous operation of systems or larger integrated systems, such as datacenters, servers in the datacenters, and houses. These systems are usually powered from an electrical grid including wall-mounted electrical outlets. Batteries are used in several scenarios. For example, when the power from the grid is out, energy from the backup battery is used for safe operation, such as data backup and safe-system shutdown. When the electricity price is high, some systems use battery energy that was charged when the electricity price was low. This is called "peak shift." Also, when systems require more power than the capability of the grid during peak, some systems use the supplemental power from the batteries in addition to the grid as shown in Figure 11.5. This ensures the performance of the systems, such as datacenters, without breaking the maximum power contract with the electricity company. This is called "peak shaving."

Lead-acid batteries have been widely used in this area because of their affordability and safety. However, the share of Li-ion batteries is increasing because of their higher energy density leading to a smaller size and footprint.

### 11.5.2 Requirements to Backup/Stationary Battery

For backup and peak shift, the battery spec needs to meet the characteristics necessary for the system operation, including the power requirement and its duration. For example, when the grid power goes out, the system may need to complete the ongoing operation, move the data in the volatile memory to the nonvolatile storage, and shutdown the system as needed. The battery needs to support all procedures, not only when the battery is fresh, but also when the battery is degraded. Furthermore, before the battery becomes too degraded to support the procedures, the user needs to replace the battery. Therefore, an IC such as a fuel gauging IC needs to continuously monitor full-charge

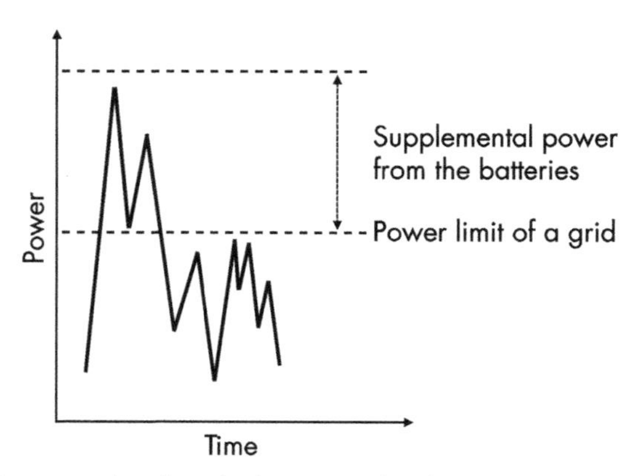


**Figure 11.5** An example of peak shaving with a battery.

capacity and impedance of the battery and communicate with the system to notify the user when the battery needs to be replaced.

For peak power shaving, the battery is used as supplemental power when the system peak power exceeds the power limit from the grid. The battery is required to support the excess power without hitting the shutdown voltage after IR drop. When the system does not need high power and part of the grid power is available, the battery is recharged. This means that the battery experiences sporadic discharging and charging many times. When the battery is at high SOC (e.g., 100% SOC), such repeated discharging and charging accelerates battery degradation more than when the battery is at low SOC, as explained in Chapter 10. Therefore, the battery SOC and charging voltage should not be unnecessarily high to extend the battery longevity. However, it should be high enough to support the peak power shaving, considering the battery voltage after IR drop, and the required energy to support the peak. Finding the optimized recharging voltage is key to successfully performing peak power shaving at low cost of ownership. When the battery is degraded, the capacity is decreased, and impedance is increased. Therefore, depending on the degradation status, it is necessary to increase the recharging voltage for continuous support of peak power shaving without hitting the shutdown voltage [4].

Because the batteries have high energy, it is also important to use proper protection circuits and meet fire codes for installation of Li-ion batteries. This is to prevent unsafe operation and thermal runaway propagation.

## 11.6 BATTERIES FOR ELECTRIC VEHICLES

### 11.6.1 EV Battery Usage and Requirements

Many EVs are powered with Li-ion batteries. The battery pack of each EV provides several ten to a hundred kWh, for example, an 85-kWh Li-ion battery pack that consists of 16 battery modules connected in series. In this example, each module is a 6S74P battery consisting of 444 cells in a 18650 form-factor. In total, 7104 cells are used, making a 96S74P pack. Prismatic or pouch form-factor cells can also be used.

To avoid over charging/discharging, multiple battery management ICs in an EV battery pack typically monitor and balance the voltages of each serial connection in addition to monitoring the overall voltage. For example, one IC monitors each voltage up to 6S, and 16 stacked ICs monitor 96S in total. The stacked ICs communicate with each other and report the status to the host in the EV.

EVs are required to operate over a wider temperature range than consumer electronics systems. For example, EVs may be exposed to the temperature range from –40°C to 85°C whereas consumer electronics typically operate from –20°C or 0°C to 60°C. EV battery systems generally include a battery thermal management system (BTMS) such as a cooling system, most using a liquid coolant circulated through a series of metal pipes to transfer heat away from the battery pack [5].

If the cost of the EV battery pack is ~$150 per kWh [6], an 85-kWh battery pack costs $12,750. A large part of the vehicle price comes from the battery cost, therefore expensive materials, such as cobalt as a $LiCoO_2$ cathode, is avoided in EV batteries. Low- or no-cobalt cathodes, such as NMC, NCA, and LFP cathodes, are popular.

EVs require high discharging current while driving at high speed and higher current during acceleration. When braking, the kinetic energy is converted into electric energy and used to recharge the battery. This is called regenerative braking. Therefore, sporadic discharging and recharging happen during EV operation. EV batteries are warranted for many years or long-driving mileage. One contract example is more than 70% capacity retention in 8 years or 100,000 miles, whichever comes first. In contrast, Li-ion batteries for consumer electronics systems typically warrant at least 80% capacity retention after 500 to 1,000 cycles at 25°C, which is much shorter than EV batteries.

### 11.6.2 Algorithms for EV Batteries

For EV batteries, charging to high SOC is typically avoided because the long warranty period needs to be supported. For example, even if the battery can technically be charged to 100 kWh from a chemistry standpoint, the battery is charged to less than 100 kWh. This is the method to extend the battery longevity, explained in Chapter 10. The lower charging limit (e.g., 90% SOC) is defined as 100% charging in the EV and 100% is shown to the user.

The driving range of EVs is largely affected by power consumption and impedance. When the air conditioner is turned on while driving, the remaining driving range decreases because more energy is consumed. Also, when a user is driving an EV fast or uphill, the remaining driving range decreases. This is because driving at high speed or uphill draws higher current from the battery. This causes more loss by IR drop and hits the shutdown voltage earlier. When the battery temperature is low, the usable capacity of the battery decreases due to higher impedance. The fuel gauging algorithm for EVs is required to safely estimate the remaining driving range considering these factors. Otherwise, the car may stall on the road.

## 11.7 KEY CONSIDERATION FOR LONGER BATTERY LIFE

In any system, longer battery life is preferred. To enable this, there are several methods:

- *Chemistry enhancement by using the cathode and/or anode materials that provide more capacity and energy.* For example, instead of Ni-MH batteries, using Li-ion batteries extends battery life. If rechargeability is not needed and high impedance is acceptable, lithium thionyl chloride ($LiSOCl_2$) batteries may be an option.
- *Better space utilization.* If a battery is changed from user-replaceable to service-replaceable, a hard battery enclosure may not be needed. This makes the battery larger and thicker, leading to higher capacity. For example, a battery in a portable system, such as smartphones, used to be user-replaceable, and a thick plastic enclosure was used around the battery for safe handling by the end users. When the battery is embedded in the system and cannot be replaced by the end users, the thick enclosure is replaced with a thin plastic film. The space for the thick enclosure is utilized to make the battery larger and thicker, resulting in higher capacity and longer battery life. Utilizing a nonrectangular battery may also be effective for better space utilization. For example, some systems may give nonrectangular space for batteries. If a nonrectangular battery is used (e.g., a round-shaped battery for a round-face smartwatch, an L-shaped battery for a complex space), dead space is mitigated, which enables longer battery life. Curved or flexible [7] batteries also provide opportunities for additional energy stor-

age, for example, a watch wristband where a curved battery may be embedded.

- *Impedance reduction.* By using a thicker current collector, thinner separator, or thinner electrode, battery impedance can be reduced. While these reduce energy density in general, impedance reduction delays hitting the shutdown voltage because of the mitigated IR drop. Balanced energy density and impedance will extend battery life.

## 11.8 SUMMARY

This chapter explained the application of Li-ion batteries to various kinds of systems. To enable longer battery life, not only battery energy density but also impedance need to be considered.

## 11.9 PROBLEM

Problem 11.1

There is a vacuum cleaner which is powered by a Li-ion battery. A genuine battery pack shows 2.0 Ah and provides 6 minutes run time. There is a compatible battery pack from a different manufacturer. The dimensions of the battery are the same. It shows 2.1 Ah which is 5% greater than the genuine battery pack. However, when used, the user noticed that the run time is 5 minutes which is 20% less. Explain a possible reason.

Answer 11.1

Energy density and impedance are typically a tradeoff. A higher energy–density battery may not always provide longer battery life because of worse (i.e., higher) impedance. In this problem set, the genuine battery may intentionally be designed with lower energy density to enable lower impedance and provide longer run time, compared to the compatible battery.

## References

[1] Kunjal, P., et al., "Requirements for Next Generation Wearable Display and Battery Technologies," *Society for Information Display International Symposium,* San Francisco, CA, Vol. 47, Issue 1, May 22–27, 2016, pp. 570–573.

[2] Matsumura, N., "Expectations and Unique Challenges of Drone Batteries," *The Battery Show 2017,* Michigan, 2017.

[3] Matsumura, N., "Selection of Battery Chemistry and Charging Algorithm for IOT Devices," *Thin Film User Group in Northern California Chapter of the American Vacuum Society,* California, 2017.

[4] Matsumura, N., et al., "Battery Charge Termination Voltage Adjustment," United States Patent No. 10,985,587.

[5] Katoch, S., et al., "A Detailed Review on Electric Vehicles Battery Thermal Management System," *IOP Conference Series: Materials Science and Engineering,* Vol. 912, 2020, p. 042005.

[6] Department of Energy (DOE), "FOTW #1206, Oct 4, 2021: DOE Estimates that Electric Vehicle Battery Pack Costs in 2021 Are 87% Lower Than in 2008," October 2021, https://www.energy.gov/eere/vehicles/articles/fotw-1206-oct-4-2021-doe-estimates-electric-vehicle-battery-pack-costs-2021.

[7] Wehner, L., et al., "Multifunctional Batteries: Flexible, Transient, and Transparent," *ACS Central Science,* Vol. 7, No. 2, 2021, pp. 231–244.

# 12

# AI/MACHINE-LEARNING/DEEP-LEARNING APPLICATION TO BATTERY CHARGING

## 12.1 INTRODUCTION

Artificial intelligence (AI), which includes machine learning (ML) and deep learning (DL), has been used everywhere in our lives, for example, internet search engine, image search, natural language processing including translation and autocomplete, and social media. The AI application has also been extended to battery algorithms, such as charging and fuel gauging. Algorithm development may be thought of as an area of software engineers. However, if battery engineers who already have battery knowledge also learn AI/ML/DL overview, creative collaboration with the software engineers is possible. For example, the battery engineer proposes an ML algorithm to extend battery longevity with chemistry knowledge and the software engineer optimizes the program with coding skill sets.

This chapter explains basic knowledge of AI/ML/DL for battery engineers and how to use it with real examples.

## 12.2 DIFFERENCE BETWEEN AI, ML, AND DL

There are several definitions for artificial intelligence, machine learning, and deep learning [1]. This book defines them as shown in Figure 12.1.

AI is any intelligent machine that does what a human brain does. For example, if the machine includes the algorithm of if-then statements and calculates something based on the input data, that is an example of AI.

ML is part of AI that learns from data (i.e., input) and makes predictions. For example, if there are datasets of $x$ and $y$ and the regression line can be drawn as $y = a + bx$ where $a$ and $b$ are constants, new $y$ can be predicted when the input $x$ is changed. The challenge is that humans need to specify what are the suitable inputs from many kinds of data. For example, the driving distance of the car at a constant speed can be predicted as a function of the driving time. However, there are many other available data, such as ambient temperature, tire pressure, and the remaining fuel. When the theory that relates the input to the prediction is known, it is easy to select the input. If the theory is not known and many kinds of data are available, many trials are needed. Input is also called a feature.

DL is the subset of ML that uses a many-layer neural network. The advantage of DL is that feature selection is not always necessary. For example, face recognition used to require feature selection by

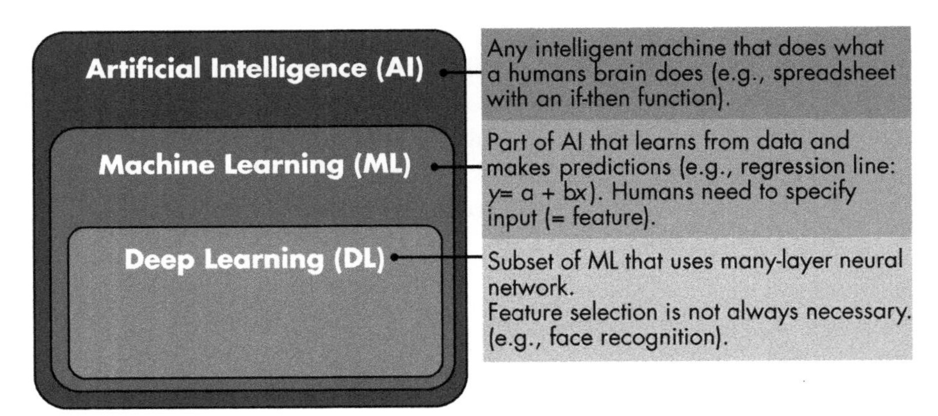


**Figure 12.1** Definition of artificial intelligence, machine learning, and deep learning.

humans, such as distances between eyes and mouth. However, with DL, humans only need to use image data and the key features hidden in the image data are automatically recognized through the multilayer neural network of DL.

The details of ML and DL are explained in the later sections.

## 12.3 PROGRAMMING ENVIRONMENT SETUP

For ML and DL, there are popular programming languages, such as Python, R, and MATLAB. The selection of the programming language may be the first challenge. The programming examples in this book are written in Python as it is popular in ML and DL. Also, in Python, there are many resources for ML and DL, such as software libraries, programming examples, and the parameters that are already optimized. However, the selection may depend on what resources are available for the user's project. For example, if there are already many resources written in MATLAB for the project, MATLAB may be used.

There are several ways to install Python. One popular and easy method is to install it through Anaconda at https://www.anaconda.com. Anaconda is the distribution platform of Python and R where the users can easily add/update/remove the libraries such as TensorFlow, which is one of the machine-learning libraries. The users can also install code editors through Anaconda, such as Visual Studio Code, referred to as VS Code, which help the users write, run, and debug programs.

On a side note, the actual implementation in the system may use a different language, such as C#. Such coding conversion is typically after the algorithm development is transferred from the battery engineering team to the software engineering team.

## 12.4 MACHINE LEARNING

This section explains how ML works. There are many models in machine learning including a simple linear model, $y = a + bx$, where $a$ and $b$ are parameters, $x$ is an input, and $y$ is a hypothesis. First, a dataset of $x$ and $y$ is split into training set and test set. The training set is used to optimize the parameters of the machine learning model and the test set is used to check the accuracy. One example is to use 80% data for training and 20% data for test.

With the training set, machine learning optimizes the parameters of the model. To do this, first it calculates an error that is the difference between the hypothesis calculated by the model and the real value. In machine learning, an error is called a cost. Machine learning calculates the costs for the training set, then optimizes the parameters so that the total cost can be minimized. After the optimization of the parameters, the accuracy of the model is checked with the test set.

### 12.4.1 ML Example: Regression Problem Case with Algebra

Figure 12.2 is an example of house prices $Y$ over house sizes $X$.

$X$ includes the data of $x_0, x_1, \ldots, x_{m-1}$. $Y$ includes the corresponding $y_0, y_1, \ldots, y_{m-1}$.

If the relationship between $X$ and $Y$ is linear as $Y = a + bX$ where $a$ and $b$ are constants, how should the regression line be drawn in Figure 12.2?

First, ML calculates the cost $J$, which is the difference between the estimated values (hypotheses) and the real values. For example, the cost is calculated as follows.

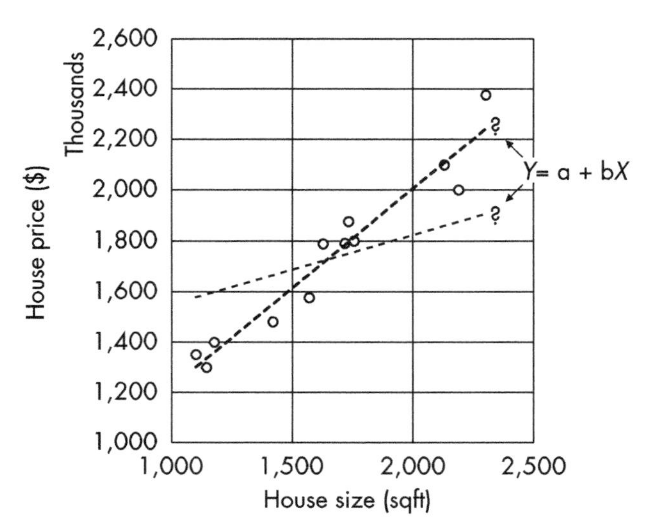


**Figure 12.2** An example of house prices Y over house sizes X.

$$J = \frac{1}{2m} \sum_{i=0}^{m-1} \left[ h(x_i) - y_i \right]^2 \tag{12.1}$$

where $m$ is the number of the data, $h(x_i)$ is the hypothesis of the house price for $x_i$ that is calculated as $a + bx_i$, and $y_i$ is the real value of the house price for $x_i$. On a side note, when the derivative is taken to optimize the equation, 2 is generated. The 2 is canceled by 1/2 in the equation.

Then, ML minimizes the cost by changing the parameters: $a$ and $b$. This is similar to the least-mean-square algorithm.

In this case, there is only one feature, $X$, as the input for the linear equation. ML can handle multiple features and nonlinear/polynomial regression models.

For faster optimization and calculation, the equation is typically vectorized in ML. For example, when there is just one data, $h(x) = a + bx$ can be written as

$$h(x) = \begin{pmatrix} 1 & x \end{pmatrix} \begin{pmatrix} a \\ b \end{pmatrix} \tag{12.2}$$

For multiple data, $h(x_i)$ can be written as

$$\begin{pmatrix} h(x_0) \\ \vdots \\ h(x_{m-1}) \end{pmatrix} = \begin{pmatrix} 1 & x_0 \\ \vdots & \vdots \\ 1 & x_{m-1} \end{pmatrix} \begin{pmatrix} a \\ b \end{pmatrix} \tag{12.3}$$

Then, the cost is calculated using (12.1).

Without vectorization, it is calculated one by one from $i = 0$ till $i = \mathrm{m} - 1$. With vectorization, the calculation is performed at once, which is faster.

Next, let's take a look at a sample data and program with Python. Table 12.1 is a sample dataset of house sizes and prices for training.

In this table, the size and price columns are the house sizes in square feet and the house prices in U.S. dollars, respectively. First, save this table as CSV (comma-delimited) and name the file as house_price_train.csv.

**Table 12.1**
Sample Dataset of House Sizes and Prices for Training

| Size | Price |
|---|---|
| 1420 | 1480000 |
| 1733 | 1875000 |
| 2131 | 2098000 |
| 2190 | 1999000 |
| 1718 | 1790000 |
| 1628 | 1788000 |
| 1144 | 1298000 |
| 1570 | 1575000 |
| 1100 | 1350000 |
| 2302 | 2375000 |

Next, Table 12.2 is a sample dataset of house sizes and prices for test.

Save this table as house_price_test.csv in the same folder with the training set.

The following is a sample ML program for the house-price estimation by the house size with linear regression. Open a code editor such as VS Code, type the following program without the line numbers, and save it as a Python file with the .py extension. The line numbers in this program are for explanation purpose in this book and are not required to type in the program. The program file needs to be in the same folder with the training and test sets.

**Table 12.2**
Sample Dataset of House Sizes and Prices for Test

| Size | Price |
|---|---|
| 1176 | 1399000 |
| 1756 | 1799000 |

*Sample ML Program*

```
 1. import pandas as pd
 2. from pandas import read_csv
 3. from sklearn.linear_model import LinearRegression
 4. df_train = read_csv('house_price_train.csv')
 5. df_test = read_csv('house_price_test.csv')
 6. print(df_train, flush=True)
 7. x = pd.DataFrame(df_train['size'])
 8. y = pd.DataFrame(df_train['price'])
 9. reg = LinearRegression()
10. reg.fit(x, y)
11. print(reg.intercept_, flush=True)
12. print(reg.coef_, flush=True)
13. print(reg.score(x, y), flush=True)
14. x_test = pd.DataFrame(df_test['size'])
15. y_test = pd.DataFrame(df_test['price'])
16. print(reg.predict(x_test), flush=True)
17. print(reg.score(x_test, y_test), flush=True)
```

An ML program such as this typically consists of the following: import of necessary libraries, data load, preparation for the ML model, such as specifying $x$ and $y$, ML model selection, training (i.e., fitting), prediction, evaluation, and test. Some additional codes such as data cleaning and graph drawing may be added depending on the user's need.

In this program, lines #1 to 3 import the necessary libraries. Pandas is one of the Python libraries for data analysis. In line #1, pandas is imported and abbreviated as pd for short and easy coding thereafter. In line #2, the module to read CSV files is imported from pandas. In line #3, from scikit-learn, which is a Python library for machine learning, a linear regression model is imported. Lines #4 and #5 read training and test sets into the data frames as df_train and df_test, respectively. Line #6 is to show the loaded training set for confirmation. This is not necessary for ML. If the test set also needs to be confirmed, print(df_test, flush=True) may be added in a new line. It is important to understand that the first-row number of the Python data is 0, not 1. For example, when the first two rows need to be selected, row 0 and 1 should be specified instead of 1 and 2. Lines #7 and #8 are the preparation to use a linear regression model in Line #9. As the linear regression model needs *x* and *y*, the corresponding columns in the training data frame are specified in lines #7 and #8. In line #9, the linear regression model is named as reg and the fitting for the model is performed in line #10. Lines #11 to #13 show the results of the fitting. When the linear regression model is expressed as $y = a + bx$, lines #11 and #12 show a and b, respectively. Line #13 shows $R^2$, which is the coefficient of determination for the linear regression. $R^2$ is one of the methods for the evaluation of the training. A different evaluation method may be used as needed. In lines #14 to #16, the linear regression model is applied to the test set. Lines #14 and #15 read the test set into x_test and y_test. In line #16, the predictions are done for the data in x_test and are displayed. In line #17, the predictions from x_test are evaluated against the true data in y_test and $R^2$ for the test set is shown. In this line also, $R^2$ is an example, and a different evaluation method may be used.

Machine learning may sound difficult to many people but is easy to use as shown above. Depending on the ML models, the code is changed. However, the structure of the program is similar. The application of the different models is explained in Sections 12.7 and 12.10.

Note that this program was confirmed to run on the versions of Python 3.9.7, VS code 1.69.1, pandas 1.4.3, and scikit-learn 1.1.1. Some modification in the program may be needed when the versions are changed. Also make sure that the necessary libraries (pandas and scikit-learn) are installed to run this program. Installation can be confirmed or performed through Anaconda.

On a side note, this program intentionally includes some lines that help readers to understand the structure of the program. Those lines may not be necessary for the people who want to know the prediction accuracy directly. If the simpler program is preferred, lines #6, #11, #12, and #16 may be removed. Also, there are many alternative codes to do something similar. For example, instead of loading train and test sets from two separate files, a train_test_split function in the sklearn.model_selection library can randomly split the combined data into training and test with the specified data size (e.g., 80% for training and 20% for test). While small-sized data is used in this section, the real cases will use a lot more data. Knowing more codes such as train_test_split will make the job easier. Also, if multiple variables are used as input, normalization is used. This transforms the features that are the inputs to the multiple variables to be on a similar scale such as the range between 0 and 1. This enables the equal weights to each variable and avoids the inappropriate influence by the features that include larger numbers. MinMaxScaler in sklearn.preprocessing is an example of a normalization code.

### 12.4.2 ML Example: Classification Problem Case

The previous section explained the ML application to the regression problem. What if the problem is binary such as yes/no? One example is the prediction of passing an exam against the study time of a student. Does the student pass the exam if the study time is 40 hours? In this case, logistic regression is used to estimate the probability where the range is between 0 and 1. There are several functions that convert the input into the probability or yes/no output. A sigmoid function is one of them and is shown in Figure 12.3.

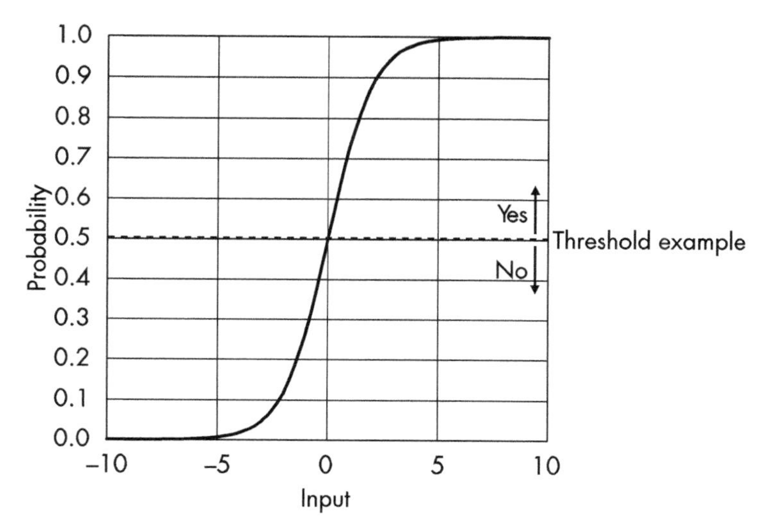


**Figure 12.3** Sigmoid function.

A sigmoid function is expressed as $1/(1 + e^{-z})$ where $z$ is the input transformed from the original input. For example, $z$ may be $a + bx$ where $a$ and $b$ are the parameters, and $x$ is the original input.

In the case of the prediction of passing an exam against study time of students, the horizontal axis is the input transformed from study time and the vertical axis is the probability of passing the exam. This gives binary answer such as yes and no by specifying the threshold in the vertical axis (e.g., 0.5) and defining the yes/no labels by the threshold such as yes (i.e., 1) when the output is >0.5 and no (i.e., 0) when the output is ≤0.5.

Similar to the regression problem, ML first calculates the cost and optimizes the parameters $a$ and $b$ to minimize the cost, using training set. The training set includes the study time of each student and the result of an exam that is pass (i.e., 1) or fail (i.e., 0).

### 12.4.3 Other ML Models

Previous sections explained linear and logistic regressions for regression problems and classification problems, respectively. There are many other ML models. Some examples are shown in Figure 12.4.

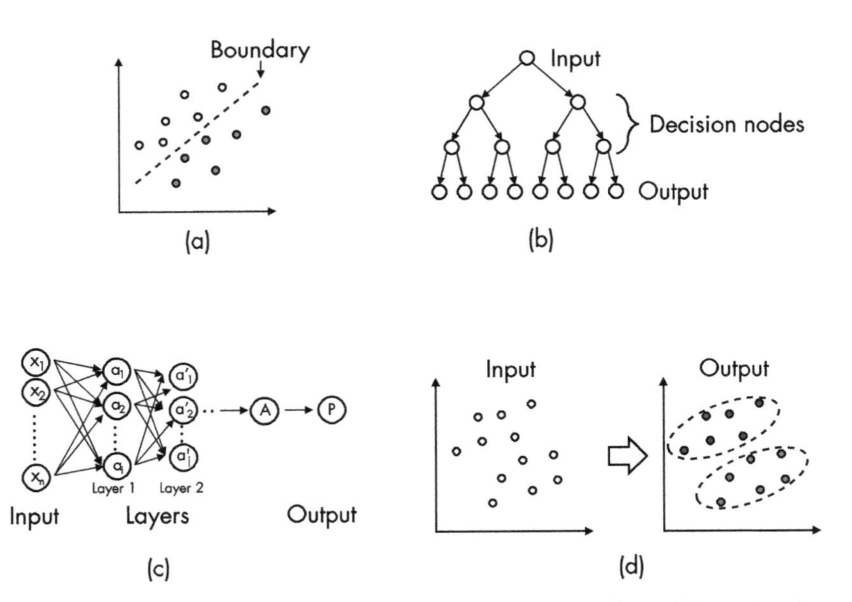


**Figure 12.4** Examples of ML models (a) support vector machine, (b) random forest, (c) neural network, and (d) k-means clustering.

For example, a support vector machine in Figure 12.4(a) is a model that finds a boundary between the two classes of data. For programming, svm model from a scikit-learn library is a code example. A random forest in Figure 12.4(b) is a model that consists of decision trees where each decision point is called a node. During training, the input features are automatically related to the output through the decision trees. The user only needs to specify the number of the decision trees in the code. The more decision trees may lead to better results although that takes longer time for training. For programming, a random forest model is available in a scikit-learn library. A neural network in Figure 12.4(c) is an algorithm that converts the input to the output through many layers of matrices. When the number of layers (i.e., depth) is more than three, it is also called deep learning [2]. Humans do not always need to specify the features as the algorithm automatically detects them during training. The details are explained in the next section.

The models explained above are trained with the data that includes the input and the known output and are called supervised learning.

There are other models that learn patterns only from the input without the known output. These models, such as k-means clustering in Figure 12.4(d), are called unsupervised learning, which is used to identify the hidden similarities or groups. For programming, the model is available in a scikit-learn library.

## 12.5 DEEP LEARNING

### 12.5.1 Neural Network and Deep Learning

For humans, input devices such as eyes and ears send the information signals (i.e., inputs) to the brain. In the brain, the connected neurons analyze the signals and may take an action (i.e., output) as shown in Figure 12.5. For example, when someone sees a dinner plate without utensils, the person may try to find utensils.

A neural network model is similar to a human brain. Figure 12.6 is an example of a neural network.

Input may be numerical data or image data that is a series of 0 and 1 in the case of binary data. The input is transformed to hidden layers that are matrices. Each connection shown as an arrow in Figure

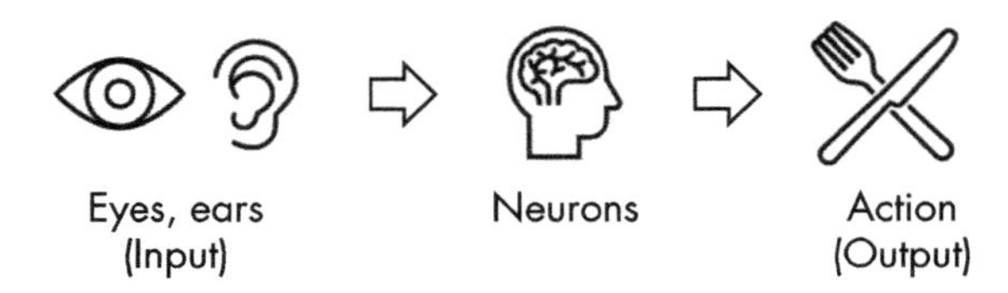


**Figure 12.5** An example of input and output through a human brain.

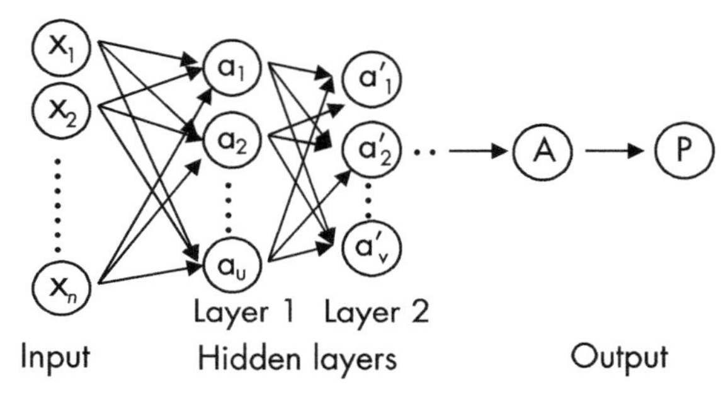


**Figure 12.6** An example of a neural network model.

12.6 is a simple calculation, such as $a = \mathrm{i} + \mathrm{j}x$ where i and j are parameters that depend on each connection, $a$ is the interim output in the hidden layer, and $x$ is the input. Such simple calculations are applied from layer to layer (e.g., from layer 1 to layer 2 where output is $a'$ and input is $a$). While each calculation is simple, there are many connections through the layers. Therefore, there are many parameters.

Through the matrices, the input is transformed to the specified dimensions of the hidden layers and the prediction is finally made. For example, when there are 1,000 images that contain the images of dogs and each image size is 400,000 bits, the input may be a 1,000 × 400,000 matrix. To predict whether each image contains a dog or not, the matrix needs to be converted to a 1,000 × 1 matrix where each row shows yes (i.e., 1) or no (i.e., 0). If there is no hidden layer, the transformation matrix contains 400,000 elements (i.e., parameters). As the neural network model contains hidden layers, there are multiple transformation matrices, therefore many parameters. During training, such parameters are optimized. When the dimensions of the matrices are higher and/or the number of the hidden layers is greater, the number of the parameters substantially increases. That requires more computational resources and time. The programmers can define and tune the numbers of dimensions and hidden layers in the program. For image processing, there are several predefined and pretrained models such as EfficientNet, ResNet, and VGG.

When there are more than three hidden layers, the neural network is called deep learning [2].

### 12.5.2 DL Applications in the Real World

While the concept of deep learning is not new, because of the recent enhancements in algorithms and computational speed, the more complex data and algorithms can be handled in a reasonable time. In 1997, Deep Blue, a chess computer with a deep-learning algorithm, won against the world chess champion [3]. In 2012, another revolution happened at the ImageNet Large Scale Visual Recognition Challenge (ILSVRC). ILSVRC is a global annual competition of software programs for image classification and detection of objects/scenes that was held from 2010 until 2017. When the competition started, the drop of the error rate was ~2% per year [4]. In 2012, the algorithm with deep learning revolutionarily dropped the error rate by ~10% [5]. Since then, the deep-learning algorithms have continued to improve the error rate and became better than the human-error rate in 2015 [4].

As we know, deep learning is used in many scenes in the world, such as image search and face detection/recognition. Deep learning can also predict the future from past input by detecting the patterns in the past through hidden layers. Long short-term memory (LSTM) is one of the models in deep learning and is used for natural language processing that predicts the next word by the previous words. It is also used for stock price prediction where the algorithm detects the hidden patterns in the past stock price changes. For programming, the models of deep learning are available in libraries such as TensorFlow.

The training and prediction may be performed in a local system such as a laptop PC, or in the cloud if they require more computational resources and time.

One important thing is that the prediction by ML, including deep learning, contains errors. The error level needs to be tolerable for implementation of the program. For example, 1% error in face recognition may be acceptable for a classification of personal photos but may not be acceptable for an alternative method to passcode.

## 12.6 TYPICAL STEPS IN ML/DL DEVELOPMENT

This section explains the typical steps in the development of ML/DL algorithms.

Step 1. Data preparation and cleaning: The data needs to be prepared and cleaned. When more data is prepared, a better prediction can be expected because the model fitting (i.e., parameter tuning) can be better with more data. The data typically includes blanks, misalignment in formats, and so forth, which need to be cleaned so that the program can recognize the data correctly. Table 12.3 is an example of battery usage data in a laptop PC.

In the laptop PCs where the Windows operating system is running, battery usage data can be retrieved by opening the command prompt and typing "powercfg -batteryreport." In the table, the first column shows the start time of the event, the second column is the power state, the third column is whether the power source is an AC

**Table 12.3**
An Example of Battery Usage Data in a Laptop PC

| **Start Time** | **State** | **Source** | **Capacity Remaining** | |
|---|---|---|---|---|
| 2022-07-0206:46:48 | Connected standby | AC | 80% | 34,304 mWh |
| 06:47:23 | Active | AC | 81% | 34,511 mWh |
| 09:45:00 | Suspended | — | 100% | 42,897 mWh |
| 09:49:02 | Active | AC | 100% | 42,897 mWh |
| 09:58:48 | Active | Battery | 100% | 42,897 mWh |
| 10:08:55 | Active | AC | 96% | 41,245 mWh |
| 10:27:11 | Connected standby | AC | 96% | 41,222 mWh |
| 10:29:42 | Connected standby | Battery | 96% | 41,222 mWh |
| 14:29:43 | Suspended | — | 94% | 40,113 mWh |
| 20:01:55 | Active | Battery | 93% | 39,697 mWh |
| 20:16:44 | Connected standby | Battery | 85% | 36,533 mWh |
| 20:18:44 | Connected standby | AC | 85% | 36,371 mWh |
| 20:18:53 | Active | AC | 85% | 36,359 mWh |
| 20:49:55 | Connected standby | AC | 100% | 43,093 mWh |
| 2022-07-0301:49:54 | Suspended | — | 97% | 41,580 mWh |
| 08:05:08 | Connected standby | Battery | 96% | 41,083 mWh |

adapter or a battery, the fourth column is the battery SOC, and the fifth column is the remaining battery energy in mWh. In the first event of the day in Table 12.3, the first column shows both date and time. However, from the second event of the day, the date is removed and only time is shown. For example, in the table, while the first event shows both 6:46 am and July 2, 2022, the second event shows only 6:47 am. To use the data in ML, the date needs to be added to all start times. Also, the timestamp of the first event is written in one cell without a space between date and time. To recognize this as date and time in ML, date and time need to be split and parsed as date and time formats, respectively. When the state is suspended, the source column is blank. If the information of the suspended state is not important, the rows may be removed. These are examples of data cleaning.

Step 2. Model selection: There are many ML models such as linear regression, logistic regression, deep learning, and others. Depending on the type of prediction and input features, several models are tested. During the model selection, the models used for a similar case may be referred.

Step 3. Training: After the model is selected, with part of the data, the parameters in the model are trained to minimize the cost. For example, randomly selected 80% data is used for training.

Step 4. Testing and evaluation: The trained models are tested with the remaining data such as 20% data, and the errors between the predictions and actual values are calculated. There are many metrics to evaluate the accuracy; for example, a simple accuracy percentage that is calculated as (the number of correct predictions)/(the number of total predictions) × 100, mean absolute error (MAE) that is calculated as:

$$(MAE) = \frac{1}{m} \sum_{i=0}^{m-1} \left| h(x_i) - y_i \right| \tag{12.4}$$

and mean squared error (MSE) that is calculated as follows:

$$(MSE) = \frac{1}{m} \sum_{i=0}^{m-1} \left[ h(x_i) - y_i \right]^2 \tag{12.5}$$

In these calculations, $m$ is the number of the data, $h(x_i)$ is the hypothesis for input $x_i$, and $y_i$ is the real value for input $x_i$.

$R^2$, the coefficient of determination for linear regression used in Section 12.4.1, is also one of the evaluation metrics.

Step 5. Hyperparameter tuning: To make the prediction more accurately, hyperparameters may be tuned depending on the model. Hyperparameters are typically defined as the parameters that affect the training of the model, for example, the number of hidden layers in deep learning, the number of decision trees in a random forest model, and the number of iterations to tune the parameters during training. After the hyperparameters are changed, training, test, and evaluation are performed again. In this step, overfitting needs to be avoided. Figure 12.7 is an example of optimized fitting and overfitting.

When the relationship between $x$ and $y$ is theoretically linear, a linear regression model is probably applicable even if the measured data includes errors. For example, Figure 12.7 shows the relationship between volume of a metal and its weight. While the measured data in Figure 12.7(a) includes some errors, linear regression would be applied as shown in Figure 12.7(b). To perfectly fit the model to the data, polynomial regression may be used as shown in Figure 12.7(c). In this case, the accuracy may be improved for the training set but would be worse when test set or new data is used. This is because polynomial regression is theoretically incorrect. Preparing a lot of data for training and evaluation is effective to check overfitting. If hyperparameters are tuned by repeatedly seeing the evaluation result with test set, prediction is biased by the test set. To avoid this, a dataset is split into

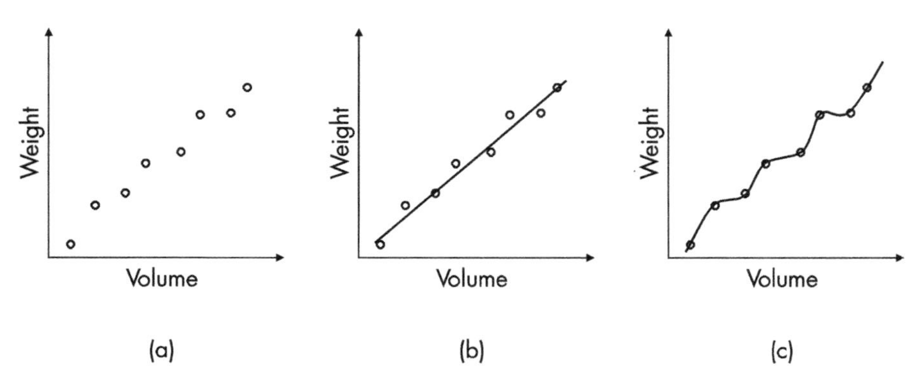


**Figure 12.7** An example of (a) data, (b) optimized fitting, and (c) overfitting.

training, validation, and test sets in advance. Hyperparameters tuning is evaluated with a validation set, then the final model is evaluated with a test set.

Step 6. Implementation: After the model with the parameters and hyperparameters is optimized and finalized, it is implemented for use. The model may be trained again in a certain period with the latest data. This is because the trend of the data may be changed.

## 12.7 CONTEXT-BASED BATTERY CHARGING: ML/DL APPLICATION TO EXTEND BATTERY LONGEVITY

### 12.7.1 Introduction

ML/DL applications to batteries have been researched in many areas, for example, battery longevity prediction from early degradation stage, battery state of health estimation, and discovery of new solid-state electrolytes [6, 7]. With ML/DL and battery degradation knowledge, it is also possible to extend battery longevity. The context-based battery charging is one of the examples [8]. Before explaining the algorithm, let's review what causes battery degradation. Some of the recent portable systems such as smartphones and laptop PCs are capable of fast charging. Such systems may always fast charge the batteries to full. However, always fast charging and/or always full charging may unnecessarily degrade the battery cycle life. Figure 12.8 is the comparison of cycle life between full and fast charging, full charging, and 80% charging.

In these cycle tests, Li-ion battery cells with $LiCoO_2$ cathode and graphite anode were used. The spec of the cell allows 4.4V charging and 3.0V discharging cutoff. In the cycle tests, discharging was always performed at 0.5C to 3.0V. In the cycle test of full and fast charging, charging condition was 1.0C CC (fast charging) until 4.4V, followed by 4.4V CV (0.05C cutoff). In the cycle test of full charging, CC charging was performed at 0.5C (normal charging speed) until 4.4V, followed by 4.4V CV (0.05C cutoff). In the cycle test of 80% charging, 0.5C CC was performed until 4.15V, followed by 4.15V CV (0.05C cutoff).

Compared to full charging, full and fast charging showed worse cycle life. This is because fast charging increased battery temperature due to joule heat and accelerated the degradation reactions. When battery charging speed was normal and charging was limited to 80%

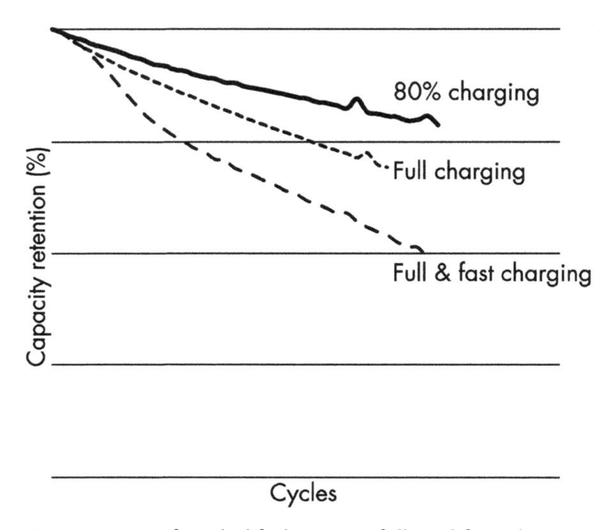


**Figure 12.8** Comparison of cycle life between full and fast charging, full charging, and 80% charging.

by lowering the charging voltage, cycle life was extended, compared to full charging. This is because battery degradation due to high voltage (i.e., high SOC) was avoided.

Avoiding full and/or fast charging not only extends cycle life, but also mitigates impedance increase due to cell degradation. This reduces IR drop during discharge, especially at peak power, and helps the system to maintain high performance at low SOC.

Unnecessary full charging and/or fast charging can be avoided when a user charges the battery as needed and fast charges the battery when needed. However, it is cumbersome if the user is required to change the charging schemes manually every time. There is a charging algorithm with ML that enables automatic adjustment of charging schemes by predicting the user's fast charging necessity and the required charging level from the user's past usage data. This is called context-based battery charging [8].

### 12.7.2 Procedure of Context-Based Battery Charging

Figure 12.9 is an example of the flowchart for context-based battery charging.

In the case of a laptop PC as an example, the context-based battery charging algorithm obtains the past battery usage data, cleans the

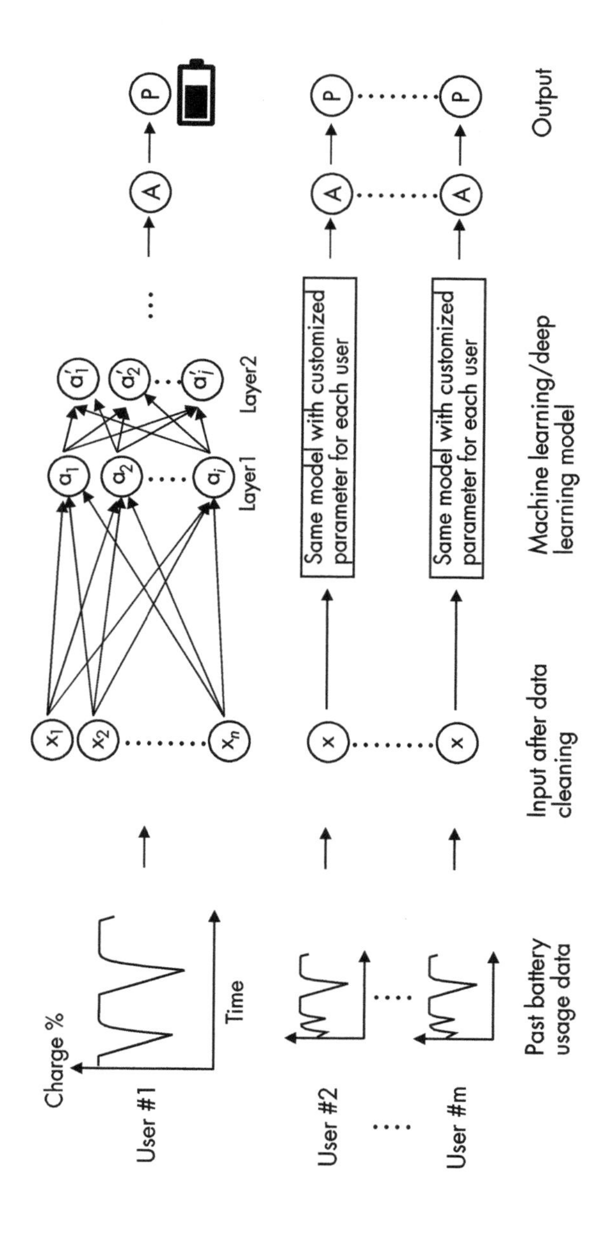


**Figure 12.9** A training and prediction flow of context-based battery charging.

data, trains the ML/DL models, and predicts the fast-charging necessity and the required charge level when an AC adapter is plugged. The ML/DL model is trained for each user with their own battery usage data. Therefore, the parameters of the model and the predictions are customized for each user.

Figure 12.10 is an example of the procedure for the context-based battery charging simulation [8].

In this work, first, the laptop PC usage data of the randomly selected 120 users was obtained. The data contains each user's battery usage for more than one year before the shelter-in-place that happened in 2020 due to the pandemic. The first 30 days of the usage data were used to train the models for each user and the remaining data was used for testing.

### 12.7.3 Results of Context-Based Battery Charging

This section introduces the results of the context-based battery charging simulation shown in Figure 12.10. First, based on the past required charging percentages, the ML algorithm predicted the next day's required charging percentage for each user. These input and output are sequential. Figure 12.11 shows the result of the necessary charging level prediction.

In this figure, the *x*-axis shows the negative error that was counted when the predicted percentage was less than the actual required percentage. The *y*-axis is the cumulative percentage of the successfully

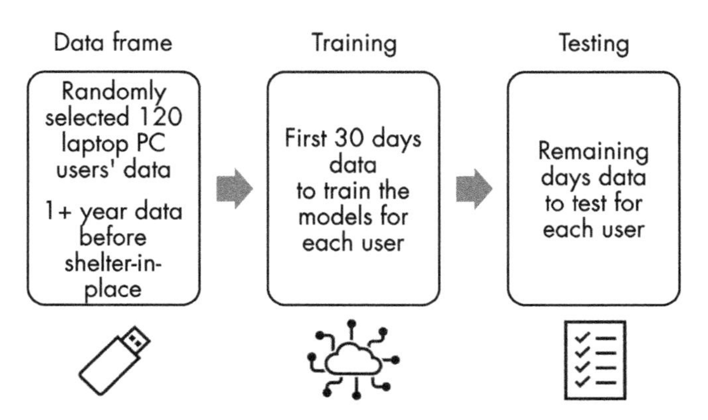


**Figure 12.10** Procedure of training and testing for the context-based battery charging simulation.

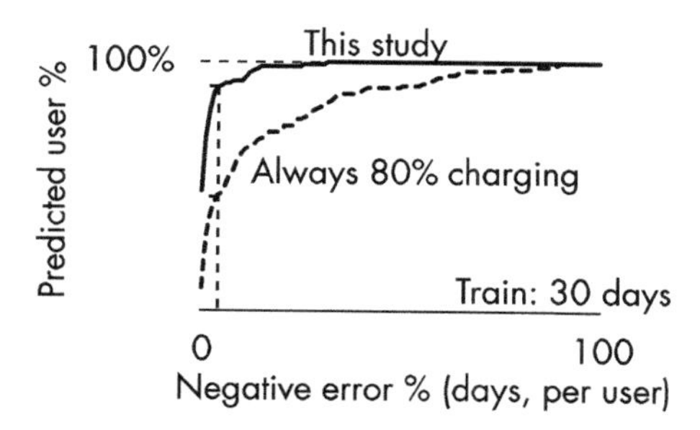


**Figure 12.11** Cumulative users percentage within negative error percentage for necessary charging level prediction.

predicted users within the negative error percentage. For example, if 100% negative error (i.e., any negative error) is allowed, 100% users are covered. Figure 12.11 shows that a large number of the users, as much as 90%, were successfully predicted within the small negative error percentage of 4%. Within 4% negative error means that the error may happen once or less over 25 days. Such an error may be because of sudden business travel, which users can avoid by temporarily disabling the context-based charging and charging the battery more in advance. It can also be avoided by considering the future schedule from the scheduling application as part of context. The predicted charging percentage was 81% on average, which means that 19% unnecessary charging was avoidable. In the case of Li-ion batteries with $LiCoO_2$ cathode and graphite anode, 81% charging is, for example, equivalent to 1.7 times better battery longevity, compared to always using 100% charging.

Some laptop PCs have software that allows users to manually limit charging to 80%. However, always limiting the system to an 80% charge may negatively affect user experience when users need more than 80% charge. Figure 12.11 also shows the cumulative percentage of users within negative error percentage when 80% charging was always performed. As the figure shows, always 80% charging covered the smaller number of the users (e.g., 47%) within the same negative error (e.g., 4%). This proves that the ML model performs better.

Next, unnecessary fast charging was predicted by the ML model every time an AC adapter was plugged in. The result is shown in Figure 12.12.

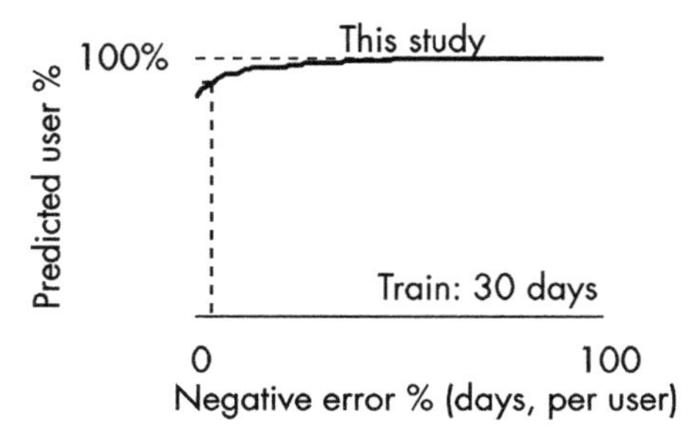


**Figure 12.12** Cumulative users percentage within negative error percentage for unnecessary fast charging prediction.

In this chart, the *x*-axis shows the negative error percentage that was counted when the prediction of fast charging unnecessity was not always true. Prediction was conducted every time an AC adapter was attached. To avoid underestimation of negative error frequency, when even one negative error happened in a day, the day was counted as a negative error day. Negative error percentage was calculated as percentage of negative error days in test period.

As Figure 12.12 shows, a large number of users, more than 91%, were successfully predicted within a small negative error of 4%. This potentially avoids 30% of unnecessary fast charging on average, which leads to, for example, 1.3 times better battery longevity in the case of Li-ion batteries with $LiCoO_2$ cathode and graphite anode.

As these results show, it is possible to predict the required charging level and fast charging necessity by ML models. In this example, the parameters customized during training were not updated during testing for an experimental purpose. If the parameters are updated continuously, context can consider the life pattern change over time and even better prediction is expected. If the context includes more information, such as the user's future schedule and location, more customized and accurate prediction is possible.

This section introduced the application of context-based battery charging for laptop PCs. It is also possible to use the same technique for electric vehicles, Internet of Things devices, backup batteries, and many other systems. The algorithms contribute not only to the longevity extension but also to sustainability because less battery replacement is required.

## 12.8 TYPICAL QUESTIONS AND ANSWERS

Here are some typical questions for ML/DL and my answers.

With ML/DL, isn't the knowledge of scientific theories needed?

- If the theory behind the data is known, the model selection and parameter fitting can be easier, faster, and more accurate. Still some ML/DL may predict something even if we don't know the theory, such as face recognition where a DL algorithm automatically finds the features. Any ML/DL model outputs something whether it is accurate or not. The key is to analyze the accuracy.

Is a spreadsheet software (e.g., Excel) enough for data analysis?

- For a two-dimensional analysis with small-sized data such as several megabytes, a spreadsheet software may be enough and easier. However, when large-sized data with multiple features needs to be handled, Python would be faster and easier.

Does a battery engineer have to be an ML expert?

- A battery engineer does not have to be a programming expert. However, understanding how ML/DL works would help to collaborate with the programming experts.

## 12.9 SUMMARY

In this chapter, we learned necessary knowledge of AI/ML/DL algorithms for battery engineering. The parameters of the algorithms are optimized during training, and the accuracy is checked during testing. As any algorithm outputs something, the accuracy analysis is important.

ML/DL applications for batteries are new areas, therefore, there are many opportunities. For example, context-based battery charging may consider more contexts such as users' future schedules and locations. Fuel gauging may be enhanced by ML prediction of battery degradation. When you face the challenges in battery engineering, ML/DL algorithms may be solutions.

## 12.10 PROBLEM

Problem 12.1

There is a smartphone user who uses the smartphone only with the battery while awake and fully charges the battery overnight every day. Table 12.4 is the user's daily battery usage data that shows the date and the necessity of more than 90% charging as >90%_charging_necessity.

In this table, when >90% charging was needed, 1 is shown. In other words, when >90% charging was not needed, 0 is shown. The dayofweek is the day of the week for the date. The smartphone runs on a Li-ion battery with $LiCoO_2$ cathode and graphite anode. If the system can predict the >90% charging necessity for the future and

**Table 12.4**
Daily Battery Usage Data of a User with >90% Charging Necessity

| Date | dayofweek | >90%_charging_necessity |
|---|---|---|
| 6/1/2022 | 3 | 1 |
| 6/2/2022 | 4 | 1 |
| 6/3/2022 | 5 | 1 |
| 6/4/2022 | 6 | 0 |
| 6/5/2022 | 7 | 0 |
| 6/6/2022 | 1 | 1 |
| 6/7/2022 | 2 | 1 |
| 6/8/2022 | 3 | 1 |
| 6/9/2022 | 4 | 1 |
| 6/10/2022 | 5 | 1 |
| 6/11/2022 | 6 | 0 |
| 6/12/2022 | 7 | 0 |
| 6/13/2022 | 1 | 1 |
| 6/14/2022 | 2 | 1 |
| 6/15/2022 | 3 | 1 |
| 6/16/2022 | 4 | 1 |

avoid >90% charging overnight when 0 is predicted, that extends battery longevity.
For the next three days after the table, predict >90%_charging_necessity using logistic regression.
In Python, a logistic regression model is available from a sklearn.linear_model library. The true values of >90%_charging_necessity are shown in Table 12.5.

Answer 12.1

This is a classification problem. The following is an example of a Python program with a logistic regression model from a sklearn.linear_model library. Before running the program, save Tables 12.4 and 12.5 as problem_train.csv and problem_test.csv, respectively. Store the files in the same location with the file of the program. The line numbers in the program are for the explanation purpose and are not required to type in the program.
Example of a Python program:

```
1. import pandas as pd
2. from pandas import read_csv
3. from sklearn.linear_model import LogisticRegression
4. df_train = read_csv('problem_train.csv')
5. df_test = read_csv('problem_test.csv')
6. x_train = pd.DataFrame(df_train['dayofweek'])
7. y_train = df_train['>90%_charging_necessity']
8. x_test = pd.DataFrame(df_test['dayofweek'])
9. model = LogisticRegression()
```

**Table 12.5**
True Values of >90% Charging Necessity for the User After the Previous Table

| Date | dayofweek | >90%_charging_necessity |
|---|---|---|
| 6/17/2022 | 5 | 1 |
| 6/18/2022 | 6 | 0 |
| 6/19/2022 | 7 | 0 |

```
10. model.fit(x_train, y_train)
11. model.predict(x_test)
```

Lines #1 to #3 import the necessary library, module, and model. Lines #4 and #5 read the training and test sets into the data frames. Lines #6 to #8 define the input (x_train) and output (y_train) for training, and the input (x_test) for test. Line #9 defines the name for the logistic regression model. Line #10 performs training (i.e., fitting) of the model. Line #11 shows the prediction for x_test, which is the next three days after the training period. When the program is run correctly, [1, 0, 0] is predicted, which means 1 for 6/17/2022, 0 for 6/18/2022, and 0 for 6/19/2022. These correspond to the true values shown in Table 12.5.

This section introduced the simplified problem where the user's >90%_charging_necessity patterns are determined by dayofweek and the logistic regression model works. For the real users, there are many factors that determine necessary charging levels. While the model may be more complicated, the prediction is possible as shown in Section 12.7. I hope this chapter helps to open the ML/DL door and make an even better world for the future.

Thank you for reading this book. I hope that you enjoyed not only battery knowledge but also practical applications to solve real life battery problems. While this book ends here, the evolution of battery technologies does not. I look forward to our journey.

## References

[1] https://newsroom.intel.com/news/many-ways-define-artificial-intelligence/#gs.o3tprz.

[2] IBM Cloud Education, "Neural Networks," August 2020, https://www.ibm.com/cloud/learn/neural-networks.

[3] Saletan, W., "Chess Bump," May 2007, https://slate.com/technology/2007/05/the-triumphant-teamwork-of-humans-and-computers.html.

[4] Langlotz, C., et al., "A Roadmap for Foundational Research on Artificial Intelligence in Medical Imaging: From the 2018 NIH/RSNA/ACR/The Academy Workshop," *Radiology*, Vol. 291, 2019, p. 190613.

[5] Krizhevsky, A., et al., "ImageNet Classification with Deep Convolutional Neural Networks," *Proceedings of the 25th International Conference on Neural Information Processing Systems - Volume 1*, Lake Tahoe, Nevada, December 3–6, 2012, pp. 1097–1105.

[6] Smith, K., et al., "Tutorial: Machine Learning and Artificial Intelligence in Batteries," *The 37th International Battery Seminar & Exhibit,* Florida, 2020.

[7] Sendek, A., et al., "Machine Learning-Assisted Discovery of Solid Li-Ion Conducting Materials," *Chemistry of Materials,* Vol. 31, No. 2, 2019, pp. 342–352.

[8] Matsumura, N., "Context-Based Battery Charging Algorithm, an Application of Machine Learning/Deep Learning to Battery Charging for Longevity Extension," *The 39th International Battery Seminar & Exhibit,* Florida, 2022.

# ABOUT THE AUTHOR

**Naoki Matsumura** is a principal engineer at Intel Corporation. He is responsible for battery algorithm development in charging and CPU turbo-boost, and new battery-chemistry enabling for mobile devices, data centers, and IoT devices. He is frequently invited to international conferences to speak on battery topics such as machine-learning/deep-learning–based battery-charging algorithms. He also teaches at San Jose State University as an adjunct faculty on batteries, where he serves as a Materials Engineering Industry Advisory Council member. Prior to that, he held battery research and development roles at Panasonic Corporation. He earned his MS in energy science from Kyoto University, and he holds many patents.

# INDEX

## C

## M



## N

T

## Artech House Power Engineering Library

Andres Carvallo, Series Editor

*Advanced Technology for Smart Buildings,* James Sinopoli

*The Advanced Smart Grid: Edge Power Driving Sustainability, Second Edition,* Andres Carvallo and John Cooper

*Applications of Energy Harvesting Technologies in Buildings,* Joseph W. Matiko and Stephen Beeby

*Battery Management Systems, Volume I: Battery Modelings,* Gregory L. Plett

*Battery Management Systems, Volume II: Equivalent-Circuit Methods,* Gregory L. Plett

*Battery Management Systems for Large Lithium Ion Battery Packs,* Davide Andrea

*Battery Power Management for Portable Devices,* Yevgen Barsukov and Jinrong Qian

*Big Data Analytics for Connected Vehicles and Smart Cities,* Bob McQueen

*Design and Analysis of Large Lithium-Ion Battery Systems,* Shriram Santhanagopalan, Kandler Smith, Jeremy Neubauer, Gi-Heon Kim, Matthew Keyser, and Ahmad Pesaran

*Designing Control Loops for Linear and Switching Power Supplies: A Tutorial Guide,* Christophe Basso

*Electric Power System Fundamentals,* Salvador Acha Daza

*Electric Systems Operations: Evolving to the Modern Grid, Second Edition,* Mani Vadari

*Energy Harvesting for Autonomous Systems,* Stephen Beeby and Neil White

*Energy IoT Architecture: From Theory to Practice,* Stuart McCafferty

*Energy Storage Technologies and Applications,* C. Michael Hoff

*GIS for Enhanced Electric Utility Performance,* Bill Meehan

*IEC 61850 Demystified,* Herbert Falk

*IEC 61850: Digitizing the Electric Power Grid,* Alexander Apostolov

*Introduction to Power Electronics,* Paul H. Chappell

*Introduction to Power Utility Communications,* Harvey Lehpamer

*IoT Technical Challenges and Solutions,* Arpan Pal and Balamuralidhar Purushothaman

*Lithium-Ion Batteries and Applications: A Practical and Comprehensive Guide to Lithium-Ion Batteries and Arrays, from Toys to Towns, Volume 1, Batteries,* Davide Andrea

*Lithium-Ion Batteries and Applications: A Practical and Comprehensive Guide to Lithium-Ion Batteries and Arrays, from Toys to Towns, Volume 2, Applications,* Davide Andrea

*Lithium-Ion Battery Failures in Consumer Electronics,* Ashish Arora, Sneha Arun Lele, Noshirwan Medora, and Shukri Souri

*Microgrid Design and Operation: Toward Smart Energy in Cities,* Federico Delfino, Renato Procopio, Mansueto Rossi, Stefano Bracco, Massimo Brignone, and Michela Robba

*Plug-in Electric Vehicle Grid Integration,* Islam Safak Bayram and Ali Tajer

*Power Grid Resiliency for Adverse Conditions,* Nicholas Abi-Samra

*Power Line Communications in Practice,* Xavier Carcelle

*Power System State Estimation,* Mukhtar Ahmad

*Practical Battery Design and Control,* Naoki Matsumura

*Renewable Energy Technologies and Resources,* Nader Anani

*Signal Processing for RF Circuit Impairment Mitigation in Wireless Communications,* Xinping Huang, Zhiwen Zhu, and Henry Leung

*The Smart Grid as An Application Development Platform,* George Koutitas and Stan McClellan

*Smart Grid Redefined: Transformation of the Electric Utility,* Mani Vadari

*Sustainable Power, Autonomous Ships, and Cleaner Energy for Shipping,* John Erik Hagen

*Synergies for Sustainable Energy,* Elvin Yüzügüllü

*A Systems Approach to Lithium-Ion Battery Management,* Phil Weicker

*Telecommunication Networks for the Smart Grid,* Alberto Sendin, Miguel A. Sanchez-Fornie, Iñigo Berganza, Javier Simon, and Iker Urrutia

*A Whole-System Approach to High-Performance Green Buildings,* David Strong and Victoria Burrows

For further information on these and other Artech House titles, including previously considered out-of-print books now available through our In-Print-Forever® (IPF®) program, contact:

Artech House
685 Canton Street
Norwood, MA 02062
Phone: 781-769-9750
Fax: 781-769-6334
e-mail: artech@artechhouse.com

Artech House
16 Sussex Street
London SW1V 4RW UK
Phone: +44 (0)20 7596-8750
Fax: +44 (0)20 7630-0166
e-mail: artech-uk@artechhouse.com

Find us on the World Wide Web at: www.artechhouse.com